HELLO, ROBOT. has been layouted by an algorithm in collaboration with Double Standards, Berlin.

Introduction

CONTENTS

DESIGN BETWEEN HUMAN AND MACHINE

HELLO, ROBOT.

What we nowadays call "robot" has been given numerous names during mankind's cultural history: Prometheus, Golem, Frankenstein's monster, cyborg or R2-D2 are just a few of them. All of these are evidence of mankind's deep-rooted desire to overcome, improve, and mirror itself with the help of technology. However, this doesn't mean that these "robots" had anything to do with our everyday lives. The twentieth century's few robots strolled around world exhibitions as lonely exhibits or worked on the assembly lines of factories. Their only way to be a part of our daily routine was wobbling around our children's rooms as tin playmates.

These circumstances have changed radically since the beginning of the twenty-first century. Be it drones, care robots, self-driving cars, intelligent algorithms, exoskeletons, or cooperating industrial robots – robot technology is becoming a part of our every day life and changing it essentially. In this, design plays a key role, as it is the designers who shape the links to the new cosmos of interaction between mankind and machines. These connections can be humanoid robots as well as cockpits of autonomous cars, the interfaces of adaptive computer programmes, or our intelligent homes' touch pads.

The exhibition *Hello, Robot. Design between Human and Machine* displays more than 150 exhibits, presenting the versatility of robotics today. At the same time, it broadens our perspective on the ethical, social, and political questions connected to the subject, because while robotics has become more accessible, personal, and often even indispensable, the question that has to be answered in every individual case is whether it improves our world. The exhibition shows that this question is as old as the cultural history of robots, and that it is up to us – the users, designers, businesspeople, and politicians – to shape robotics' entry into our daily life in a responsible and beneficial manner.

I am sincerely grateful to the two museums that organised and produced this ambitious exhibition with us: the MAK – Austrian Museum for Applied Art / Contemporary Art and the Design museum Gent. I would also like to thank the German Federal Cultural Foundation as well as the Global Sponsor ABB, which both generously supported this project. Last but not least, my thanks go to the curators Amelie Klein, Thomas Geisler, Marlies Wirth, and Fredo De Smet. They approached the topic by asking subtle and often ambiguous questions instead of being content with simple answers. This was a smart move, for isn't the ability to ask the right questions mankind's last tiny lead over machines?

MATEO KRIES
DIRECTOR
VITRA DESIGN MUSEUM

CHRISTOPH THUN-HOHENSTEIN
GENERAL DIRECTOR
MAK – AUSTRIAN MUSEUM OF APPLIED ARTS / CONTEMPORARY ART

MAKING THE WORLD A BETTER PLACE WITH ROBOTS

The smartphone first saw the light of day in 2007, but before long it developed into a super-robot which half of the world's population now carries in their pockets as an indispensable accessory. 2007 thus marks the beginning of a new digital modernity influenced by robotics. Neither humans nor digital machines today can precisely predict how our world will work in twenty years. It is clear, however, that robots will play the leading part; they will not only outnumber humans, but they will be capable of replicating themselves.

The idea of robots dominating our lives has long been a cause of fascination and fear – not infrequently at the same time. But the Smart New World already has us in its grasp, even if we cannot yet comprehend the full scope of the transformative power of the innovations we are being given in incremental doses. We will come to realise that robots are not necessarily humanoid in form, but that they can occur in a variety of shapes and may even be entirely invisible. They have no need of arms and legs, they will be everywhere, and constantly connect us with the whole world. The extent to which this ubiquity is seen as either beneficial and liberating or detrimental and threatening will fundamentally determine our quality of life. How can design help with this?

The starting point for the exhibition *Hello, Robot. Design between Human and Machine* is design's aspiration to serve as a mediator between humans and digital machines – a task of essential importance in the digital modernity. Design will advance to become a key creative discipline in our civilisation's further development, because robotics will change everything – how we communicate and interact with one another, how we work and relax, and how we care for ourselves and others. Robotics opens up possibilities in areas such as health care that were unimaginable only a few years ago, but it also brings with it enormous risks. Combined with specialised artificial intelligence, robots will increasingly push us out of our jobs, but the degree to which they can create new ones will depend on us: success in tomorrow's world will be predicated on not trying to fight back against the robotic tide, but rather working creatively and sustainably with the robots in the service of the common good. However, robots in the form of an already growing artificial general intelligence could bring about the end of human existence if we are unable to design a future super intelligence that is enduringly human.

Robotics in its widest sense, therefore, is *the* future topic of the twenty-first century. On the one hand, the key challenge of design is to convince us not only to respect robots (and perhaps even love them), but also to actively use them to bring about positive change. On the other hand, design must seek to turn robots into humanists. We must grasp the advance of robotics as a unique opportunity to correct the detrimental effects of an unbridled, often avaricious economic system that ruthlessly exploits the environment and to ensure the sustainable welfare of the people of our planet. If we want smart cities that are liveable, they must be both: benevolent robots and commons maintained by dedicated human effort.

On behalf of the MAK I would like to thank the Vitra Design Museum and the Design museum Gent for their stimulating collaboration and the curators of the exhibition for their intriguing concept. I wish this most important project the greatest possible success.

KATRIEN LAPORTE
DIRECTOR
DESIGN MUSEUM GENT

"HELLO, ROBOT." THE START OF A CONVERSATION.

A design heritage museum that makes an exhibition about the future. This sounds just as strange as developing robots that can take over our jobs. Welcome to the paradoxical world of today, a world which revolves around mankind, yet a world we can barely control. A world where grey is made up of thick, alternating bars of black and white. Where standing still is more exhausting than taking action. Welcome to *Hello, Robot. Design between Human and Machine.*

Although the exhibition features vintage posters, films, and toys, it is not simply about heritage. Although artists and architects are involved, it is not simply about art and culture. Although the objects were selected for the quality of their design and although you can see robots in action in the exhibition, *Hello, Robot.* is not simply about design or technology.

More than robots, heritage, art, or culture, this exhibition is an exploration of *interaction design*. You read it in the title, hear it in the curators' questions; you feel it in the design, you experience it through the installations. But more than anything, *Hello, Robot. Design between Human and Machine* seeks to initiate a conversation – a conversation that goes beyond engaging with the individual visitor.

After all, with this exhibition we aim to question the relationship between human beings and their world. In this sense we clearly and intentionally embrace our social role, because we believe that the museum is much more than a place where people come to experience beauty or wonder; a museum can also be a place for reflection, encounters, and social engagement. The museum as a driving power for change: is that naive?

Design museum Gent occupies a unique position in the city of Gent and in Flanders. Situated in the very heart of the historic city, partly housed in a magnificent eighteenth-century building, it is the custodian of a collection that was founded in 1903. With this collection, which focuses on design from 1860 to the present day, we are the only design museum in Belgium able to reflect on the relevance of design through the ages. Furthermore, the city of Gent is a rich ecosystem of designers, artists, creative entrepreneurs, technology companies, and research institutions.

We work closely with all these partners. The challenges posed are considerable. Actions are increasingly digital, technology is becoming ever more invisible. These are the objects and processes we cling to. And we do cling to them in a world in which change is the only certainty. This is why we, as a museum, must not miss the boat of technological progress, the "ro-boat" so to speak. And why we, as an open institution, help to shape our city and our world, in synergy with designers. That is our mission.

This is why, on behalf of Design museum Gent, I believe in the importance of this topic. This is why, together with Vitra Design Museum and MAK, we are investing in this ambitious exhibition. This is why *Hello, Robot.*

HORTENSIA VÖLCKERS
BOARD / ARTISTIC DIRECTOR

ALEXANDER FARENHOLTZ
BOARD / ADMINISTRATIVE
DIRECTOR

"A robot may not injure a human being. [...] A robot must obey the orders given it by human beings. [...] A robot must protect its own existence as long as such protection does not conflict with the First or Second Laws." These were the Three Laws of Robotics formulated by the Russian science fiction author Isaac Asimov in 1942. Asimov was convinced that intelligent machines would not only coexist with human beings in the future, but would even become superior to humans; hence the need for such laws. Today it appears as though humans, for the first time in history, have the technological know-how and tools to turn this science fiction vision into reality: the first driverless cars are already being tested on our motorways, stock market transactions are being brokered by algorithms, androids welcome us to hotels, and the Internet of Things is turning entire cities into intelligent machines. Robots have emerged from the factories of industrial mass production to interact with us more and more in our everyday lives. This, of course, raises questions about how we should handle the increasingly more intelligent and more autonomous world of self-learning objects. Design would seem to be a particularly suitable medium for exploring the possibilities and the limits of this new (robotic) ecosystem – as an intermediary between the unknown and the familiar and as an experimental space for the many different kinds of interactions between human and machine.

The exhibition project *Hello, Robot.*, a co-production of Vitra Design Museum, the Austrian Museum of Applied Arts / Contemporary Art (MAK), and Design museum Gent, will sharpen our "sense of the future" by directing our attention to familiar and unfamiliar forms of robotics and showing us how they are changing our lives. The German Federal Cultural Foundation would like to thank the participating institutions and the team of curators, Amelie Klein, Thomas Geisler, Marlies Wirth, and Fredo De Smet, for the realisation of a project that invites you to embark on a journey through the world of robots of today and tomorrow.

A WORD OF WELCOME FROM THE GERMAN FEDERAL CULTURAL FOUNDATION

WELCOME TO OUR WORLD

ULRICH SPIESSHOFER
CEO ABB

As the pioneering technology leader and inventor of industrial robots and their foremost manufacturer, ABB is passionate about robots. We share the widespread fascination they generate in both new and traditional fields; thus, we are proud to be a partner of the exhibition *Hello, Robot.*

The exhibition is well timed. Until recently, robots have been hidden from public view, welding metal sheets, painting car bodies, or packing food in factories. They have been kept apart from workers, isolated in metal cages for the safety of employees.

Not anymore. In 2015, ABB launched the world's first truly collaborative safe robot, YuMi, a small, lightweight machine with two arms. It is precise enough to thread a needle and intelligent enough to self-learn whilst stopping the moment it meets resistance. With YuMi, a revolution has begun that will see robots and people truly working side by side.

Design plays an essential role in helping people accept their new robotic partners. ABB put as much thought into designing YuMi for its task as into its technological development, winning the prestigious Red Dot Award for design as a result.

Over the past forty years, robots have been key to elevating the nature of work, to making industrial goods affordable and improving production quality, and to boosting prosperity and growth. As we welcome them into our daily lives, robots will bring further benefits, including improving goods even more and offering valuable assistance to people of all ages. We will see entirely new applications of robotics, and they will become as ubiquitous as mobile phones and computers, i.e. the partners of people and industry in daily life.

THE SEARCH FOR QUESTIONS

We can assume that Jacques Tati wouldn't be on Facebook were he alive today. And with their unquenchable thirst for user data, the likes of Google, Apple, Microsoft, and Amazon would hardly have been able to win the favour of the French filmmaker, who died in 1982, either. After all, Tati's unforgettable works such as *Mon Oncle* and *Playtime* made it abundantly clear what he thought of the new technology of the mid-twentieth century: not much. In one legendary scene, Tati's cinematic alter ego, Monsieur Hulot, enters his sister's fully automated kitchen.[1] First he burns his finger on a heating element, then he finds it impossible to open the kitchen cabinet. He pushes buttons and everything begins to buzz and beep. The door suddenly flies open and out rolls a jug, which falls to the floor. But nothing happens, for the jug is made of an elastic material. Relieved, Monsieur Hulot bounces it off the ground a couple of times. Then he tries the same thing with a glass. Crash! All he wanted was some iced tea.

He never says a word, but it's written clearly on his face: What is this good for? Why do we need it? Faced with the digitisation of our lives driven by companies such as the "Frightful Five of the tech industry" mentioned at the beginning of this essay,[2] we are still asking this question today and it still causes controversy. But actually it isn't a question at all, for just like in the past, technology cannot be stopped as long as it sufficiently indulges our existing habits and makes our lives easier. "Convenience is a world power", says author and Internet expert Sascha Lobo,[3] the best example of this being, of course, the smartphone. No one seemed to need a smartphone until the introduction of the iPhone in 2007, but less than a decade later it is impossible for most of us to imagine everyday life without these smart little helpers. Of course, Jacques Tati knew that progress had to progress, regardless of whether he liked it or not. "In the fully automated kitchen in *Mon Oncle,* he is not just running up against the often invoked 'malice' of the inanimate object," writes film critic Roland Mörchen, "rather he is spoofing the spirit (or rather demon) of a 'new artificiality'. *Mon Oncle* is the friendly wink of a man who knows he cannot do away with what is known as modernity."[4] And so we can be sure that, were he alive today, Tati would not be on Facebook, but he would almost certainly own a smartphone.

AMELIE KLEIN

1 See the work description for *Mon Oncle,* p. 170.
2 Farhad Manjoo, "Tech's 'Frightful 5' Will Dominate Digital Life for Foreseeable Future", in *The New York Times* (20 January 2016), http://www.nytimes.com/2016/01/21/technology/techs-frightful-5-will-dominate-digital-life-for-foreseeable-future.html, accessed on 4 December 2016.
3 Sascha Lobo, "Bequemlichkeit schlägt Datensparsamkeit", in *Spiegel Online* (September 28, 2016), http://www.spiegel.de/netzwelt/web/zugriff-auf-daten-bequemlichkeit-schlaegt-sicherheit-kolumne-a-1114091.html, accessed on 4 December 2016.
4 Roland Mörchen, "Die Anarchie der leisen Töne. Jacques Tatis pointierte Alltagskomik", in *Film Dienst* (no. 21, 1998).

5 Bruce Sterling in an interview with Amelie Klein (Turin, 19 April 2016).
6 See the work description for *R.U.R. Rossum's Universal Robot,* p. 42.
7 Carlo Ratti in an interview with Amelie Klein (Weil am Rhein, 4 July 2016).
8 Erica Palmerini, Federico Azzarri, et al., *RoboLaw – Regulating Emerging Robotic Technologies in Europe: Robotics Facing Law and Ethics,* http://www.robolaw.eu/ RoboLaw_files/documents/ robolaw_d6.2_guidelinesregulatin-grobotics_20140922.pdf, p. 15, accessed on 4 December 2016.
9 Boston Dynamics, *Atlas – The Next Generation,* on YouTube, https://www.youtube.com/ watch?v=rVlhMGQgDkY, accessed on 4 December 2016.

JUST WHAT IS A ROBOT?

The appearance of the robot in our everyday lives is just as unavoidable – its visible appearance that is, for in fact robots have been lurking in parts of washing machines, automobiles, and automatic cash dispensers for years. Of course, such creatures will not look like robots, or rather they will not take the form that most of us have come to expect. "Robots are tools for dramatic effect. They are not a piece of technology," says Bruce Sterling, science fiction author and advisor to the exhibition *Hello, Robot. Design between Human and Machine.*[5] It is no coincidence that the word "robot" is the invention of a playwright. Karel Čapek's 1920 play described a mechanical working class – in other words, a class that has been dehumanised and hence robbed of its dignity – which first rises up against its masters, human beings, before revealing itself to be the morally and ethically superior species.[6] Čapek, a staunch antifascist, was engaging in a piece of social criticism which, based on humanity's age-old desire to reproduce itself, has been expressed time and again: the robot that serves us – and the robot that destroys us …

Thus, popular culture has influenced our expectations regarding robots for almost a hundred years. They should be humanoid in form, i.e., look just like us, and they should think, communicate, and move as we do. Our fascination for these human machines has reached the world's robotics laboratories, where researchers are eagerly working on creating humanoid robots. But they really ought to know better, for at present robots are not even capable of mastering the things that humans can do only two years after they are born: walk more or less confidently on two legs, even managing to stay upright on uneven ground, stairs, ice, and sand. It's no wonder that we always find real robots a bit of a let-down when we see them. They are even worse than Arnold Schwarzenegger in *Terminator.*

11

What we often forget, however, is that robots – unlike humans – don't actually need their own enclosed bodies. They only need three things, says Carlo Ratti, director of MIT's Sense*able* City Lab and also an advisor to *Hello, Robot.*: sensors, intelligence, and actuators.[7] In other words, they require measuring instruments; software that is capable of making sense of and using the information these gather, such as light, sound, or heat; and devices that trigger a measurable physical reaction. Viewed in this light, this means any house and any environment can be a robot. A robot can observe us through numerous cameras simultaneously and, for example, regulate a city's traffic lights or adjust the lights in our living room according to what it sees. We could also describe the smartphone as a kind of mini-robot – and paired with us we could say it forms a (partially) robotic system.

Ratti's definition of a robot is certainly very broad, but it nonetheless leaves out certain things that we think of as typical characteristics of robots.[8] For example, they are supposed to teach and steer themselves, they should make autonomous decisions, and they should be at least partially physical in nature. But this is not true of every robot. Classical industrial robots can only perform the movements they have been programmed to perform; they do not make decisions on their own, nor do they learn. Surgical robots are remote controlled – mercifully – and the same is true of most drones. And the Internet is teeming with softbots, self-learning software which can chat with users or provide shopping tips, but that have no physical form. It appears that there is no universally acceptable definition of robots. Only one thing seems to be clear: yes, two-legged humanoid robots such as Boston Dynamics' *Atlas,* which over nineteen million viewers have watched stumble through the snow on YouTube, do indeed exist.[9] But robots are much more than that. They make our physical world intelligent. They transform objects into "smart objects". They can give rise to a scenario in which all of the things we know from the Internet can step out of the screen and permeate three-dimensional space.

10 Carlo Ratti in an interview with Amelie Klein (Weil am Rhein, 4 July 2016).
11 László Moholy-Nagy, *Sehen in Bewegung.* Edition Bauhaus 39 (Leipzig, Spector Books, 2014), p. 42.
12 Ibid.
13 Nicolas Nova (Near Future Laboratory), Nancy Kwon, Katie Miyake, Walt Chiu (Art Center College of Design), *Curious Rituals,* https://curiousrituals.wordpress.com/, accessed on 4 December 2016.
14 Ibid.

Nicolas Nova (Near Future Laboratory), Nancy Kwon, Katie Miyake, Walt Chiu (Art Center College of Design). *A Digital Tomorrow,* 2012. Video, 9 min 36 sec, produced as part of the study *Curious Rituals,* July–August 2012 © Nicolas Nova, Nancy Kwon, Katie Miyake, and Walt Chiu

The exhibition *Hello, Robot.* traces the successive development of our definition of the robot, as does this book. First, we encounter more or less friendly humanoid robots (as well as a vacuum cleaner) before moving on to examine robots from the spheres of work and industry. Taking a closer look, we confront the machines face to face: as smart assistants and assiduous helpers that help care for us. Finally, we ourselves meld with the robot: prosthetics and implanted chips bring the robot inside us, while robotic architecture and environments bring us inside the robot. On page 32 and at the entrance to the exhibition you will find our attempt at a robot taxonomy. It is nothing more than an incomplete approximation, for robots are just as diverse as the world they increasingly populate.

AND WHAT IS THE ROLE OF DESIGN?

If we follow the broad understanding of robots described above, this would mean that many robots are not different in appearance from non-robotic objects, such as ordinary dolls, cars, or houses, but only in how they behave. "The medieval city remains a medieval city," explains Carlo Ratti, a native of Turin, "what changes is how we interact with it."[10] Like in all other parts of the digital sphere, it is not only a question of the design of form and function, but of interaction, relationship, and the combination of the two: experience. This might sound new, but it isn't new at all. As early as 1947, László Moholy-Nagy, one of the most important figures of the Bauhaus, wrote: "Design is a complex and demanding task. It entails the integration of technological, social, and economic requirements, biological demands, and the psychophysical effects of materials, shape, colour, volume, and space: it is about thinking in relationships."[11] He continues: "There is design in the structure of emotional experiences, in family life, in work relationships, in urban planning, in cooperation among civilised people. Ultimately, all of the problems of design come together to form one large problem: 'designing for life'."[12]

How then are our interactions and relationships with the intelligent objects that increasingly surround us designed? Beyond the traditional interfaces of buttons, switches, and joysticks there are also a number of unusual gestures one is forced to perform when interacting with technology. We swipe our hands through the air when we want to open train doors and our fingers over the screen when we want to read our emails. We wave at the motion detectors when we find ourselves in darkened lavatories after making the mistake of sitting too long and we open the electronic entrance to the office with a saucy swing of the hips when we are too lazy to fish our ID cards out of our pockets. *Curious Rituals* is the name of a study conducted by Nicolas Nova, Nancy Kwon, Katie Miyake, and Walt Chiu as part of their degree course at the Art Center College of Design in Pasadena, California, which examined these and other gestural interactions with technology.[13] Their study also included a video, *A Digital Tomorrow,* which shows that things won't get any better in the future.[14] Smart devices are charged by swinging them in circles through the air, a slap on the cheek ensures better concentration when synching brainwaves, and voice recognition works just as poorly as it does today.

15 David Rose, *Enchanted Objects: Innovation, Design, and the Future of Technology* (New York, Scribner, 2015).
16 David Rose, "Enchanted Objects", TEDxBeaconStreet (16 November, 2014), https://www.youtube.com/watch?v=I_Ahhhc-ceXk, 12:52 Min., accessed on 4 December 2016.
17 See work description for *Uninvited Guests,* p. 88.
18 http://www.vitality.net/, accessed on 4 December 2016.
19 David Rose, "Enchanted Objects", TEDxBeaconStreet (16 November 2014), https://www.youtube.com/watch?v=I_Ahhhc-ceXk, 08:39 min., accessed on 4 December 2016.
20 Jo Bager, "Der Datenkrake: Google und der Datenschutz", in *c't* (10/2006), p. 168, https://web.archive.org/web/20060613011608/http://www.heise.de/ct/06/10/168/, accessed on 4 December 2016.
21 https://de.wikipedia.org/wiki/Datenkrake, accessed on 4 December 2016.
22 Wolfgang Uchatius, "Warum glaubt Google, mein Kaninchen frisst Hundefutter", in *Die Zeit* (no. 47, 10 November 2016), p. 18.

Indeed, we continue to imagine that in the future technology will always work perfectly. This is surprising, for there is nothing in the present that might indicate that this will be the case. Just how often, for example, have you spoken on the phone with your IT consultant or Internet service provider over the past month? We also tend to think that technology generally will (inter)act in our best interests – at least when it isn't focused on world domination and our ultimate destruction. What we are seeing even today, however, is a kind of well-intentioned paternalism. David Rose, researcher at the MIT Media Lab, entrepreneur, and expert for human-computer interactions, has developed a series of *Enchanted Objects,*[15] as he calls them: smart networked objects capable of fulfilling our wishes like in a fairy tale. One of these, a waste bin, doesn't just automatically order online the things we have thrown away; it also comments on the owner's consumption habits. It asks, for example: "Do you really want to order Asian mineral water again? Why don't you buy locally!?" Or reminds us: "That was your third packet of biscuits today." At least you can give the bin a kick when you're fed up with its remarks – it understands that, too.[16]

A project by the design studio Superflux offers a take on the same theme: *Uninvited Guests.*[17] In the video we are introduced to Thomas, a seventy-year-old widower who has received an assortment of smart objects from his concerned children. They are intended to help him get safely and healthily through everyday life. On the first day, Thomas reluctantly follows the ever more pestering instructions from his intelligent devices; on day two he simply ignores them. But everything is networked with everything else, and so it is that Thomas receives the first worried messages from his children: "Hi Dad, I see you're not using the smart cane today. Hope all is ok? xxx Gina." Design not only shapes our interactions with machines, it seems, but also how we interact with one another.

Superflux sees itself as a design studio that seeks a critical examination of new technologies and their effects on the world. *Uninvited Guests* is a speculative project that is meant to spur discussion. David Rose, however, has developed a smart screw cap for pill bottles that has enjoyed high levels of sales for several years.[18] *GlowCap,* the name of this intelligent device, reminds users to take their medicine. If they neglect to do so, the screw cap starts to blink, by all means a sensible reminder, for it is certainly important that patients take their medicines according to schedule. In 2010, it won the American Medical Design Excellence Award. But *GlowCap* goes one step further: if the patient fails to take his or her medicine after the reminder, the smart cap sends a message to their loved ones. And another one to the doctor. And another to the health insurance company, for they are the main distributors of *GlowCap.*[19]

JUST WHO'S THE BAD GUY HERE?

The boundaries between well-intentioned concern, surveillance, and outright espionage are blurry. In 2006, the German computer magazine *c't* referred to Google as a "Data Kraken",[20] and ever since then the term has become a byword for notorious data collectors and even had its own entry on the German Wikipedia website. According to the Wikipedia definition, Data Kraken are "systems and organisations that evaluate personal information on a grand scale and/or redirect it to third parties. In doing so, they allegedly or demonstrably are in breach of data privacy regulations or violate the personal rights postulated by privacy groups that go beyond these."[21] And even if Big Data has yet to evolve into "Smart Data", as an article in the weekly *Die Zeit* has claimed – that is, if the data collectors have not yet learned to properly classify all the information they gather[22] – it would still be naïve to believe that a health insurance company would not allow a patient who neglects to take his medication go unpunished. And if health insurance premiums are raised because a patient forgets his medicine, then we're just a hop, skip, and a jump away from a scenario in which premiums are raised on those who occasionally have one too many at the pub or dine too often at the corner chip shop.

13

The Internet unremittingly collects data about our behaviour. And with robotics, the arrival of the Internet in three-dimensional space, this is set to explode exponentially. The Internet of Things and the Smart City, all of these are projects for major corporations, and not only those that make these infrastructures available, but also those who are keen to evaluate the data we generate or sell it on to third parties like the advertising industry. "An Internet of Things," writes Bruce Sterling, "is not a consumer society. It's a materialised network society. It's like a Google or Facebook writ large in the landscape. Google and Facebook don't have 'users' or 'customers'. Instead, they have participants under machine surveillance, whose activities are algorithmically combined within Big Data silos."[23]

In an essay appearing in this book, the philosopher Rosi Braidotti speaks at length about the economisation of people. "But this exploitation is not limited to people: In substance, advanced capitalism both invests in and profits from the scientific and economic control and the commodification of all that lives. [...] Seeds, plants, animals, and bacteria fit into this logic of insatiable consumption alongside various specimens of humanity. The uniqueness of *Anthropos* is intrinsically and explicitly displaced by this equation."[24] Thomas Vašek, editor in chief of the philosophy magazine *Hohe Luft,* also introduces machines to this observation: "All of us – humans as well as robots, smartphones, and artificial intelligences of every kind – are slaves of digital capitalism. We all produce data that is economically exploitable for Google and the like, we all leave data trails in the infosphere, we are all digitally predictable – and therefore we can be easily controlled by a digital mega-superintelligence. We call it the capitalist system."[25] Before the filthy lucre we are all the same.

Unfortunately, design is all too willing to serve the will of this mega-superintelligence. But this need not be the case. Indeed, it shouldn't be the case. Even for Walther Gropius, design and ethics were inseparable. In his 1925 "Principles of Bauhaus Production", Gropius called for a "resolute affirmation of the living environment of machines and vehicles" and in doing so was clearly making a social claim: "The creation of standard types for all practical commodities of everyday use is a social necessity. On the whole, the necessities of life are the same for the majority of people. The home and its furnishings are mass consumer goods, and their design is more a matter of reason than a matter of passion."[26] In 1963, in the middle of an economic boom, the British graphic designer, photographer, and author Ken Garland published a manifesto titled "First Things First", in which he called for designers to dedicate their talents and attentions not only to the large corporations, but to socially relevant topics. A list of alternatives to advertising for cat food and striped toothpaste was followed by the statement: "We do not advocate the abolition of high pressure consumer advertising: this is not feasible. Nor do we want to take any of the fun out of life. But we are proposing a reversal of priorities in favour of the more useful and more lasting forms of communication."[27]

"First Things First" struck a chord that continues to resonate to this day. Garland's manifesto does not call into question the underlying political and economic system: "This is not feasible." After all, design is not a "neutral, value-free process", explains Katherine McCoy, a graphic artist and lecturer for two decades at the Cranbrook Academy of Art, one of the most recognised academies for design in the United States.[28] The fundamental decision of whether or not a designer offers his or her talent in the service of a Data Kraken is a political one and should be discussed as such. Perhaps this is why the "IoT Design Manifesto 1.0",[29] a ten-point list of demands concerning the design of the Internet of Things, leaves us with such an unpleasant aftertaste. Five of the ten demands are dedicated to the issues of security and data protection, which is a good thing of course. Point four reads as follows: "We keep everyone and everything secure" – a reference to attacks from hackers and similar threats. Why, then, did the manifesto's authors put this point first: "We don't believe the hype. We pledge to be sceptical of the cult of the new – just slapping the Internet onto a product isn't the answer. Monetising only through connectivity rarely guarantees sustainable commercial success."

23 Bruce Sterling, *The Epic Struggle of the Internet of Things* (London, Moscow, Strelka Press, 2014).
24 See Rosi Braidotti, "Becoming-World Together: On the Crisis of Human", p. 238.
25 Thomas Vašek, "Befreit die Roboter!", in *Hohe Luft_spezial Digitalisierung / Hohe Luft* (no. 1, 2017), p. 6.
26 Walter Gropius, "Principles of Bauhaus Production", in *Programs and Manifestoes in 20th-century Architecture* (Cambridge, MIT Press, 1970), pp. 95–96.
27 Ken Garland, "First Things First", facsimile in *Design Is History,* http://www.designishistory.com/1960/first-things-first/, accessed on 4 December 2016.
28 Rick Poynor, "First Things First Revisited", in *Emigré* (no. 51,1999), http://www.emigre.com/Editorial.php?sect=1&id=13, accessed on 4 December 2016.
29 www.iotmanifesto.com, accessed on 4 December 2016.

The issue here is not the fact that designers wish to develop a sustainable means of earning money. Quite the opposite, in fact, for there are simply too many designers who have to live in precarious conditions because their work is insufficiently valued. The problem here is that commercial success appears in the very first point of a manifesto that claims to serve as a "code of behaviour" for those involved in the development of the Internet of Things.

When "First Things First" was revised and republished at the turn of the millennium with the new title "First Things First Manifesto 2000", it received an important addition. It now reads: "We propose a reversal of priorities in favour of more useful, lasting, and democratic forms of communication – a mindshift away from product marketing and toward the exploration and production of a new kind of meaning. The scope of debate is shrinking; it must expand. Consumerism is running uncontested; it must be challenged by other perspectives expressed, in part, through the visual languages and resources of design."[30]

Bruce Sterling adds his own take: "Rather than thinking outside the box – which was almost always a money box, quite frankly – we surely need a better understanding of boxes."[31] In other words, we have to change the parameters, redefine the context, and ask different questions. "Instead of pursuing projects, defining goals, and thus describing a linear path to a solution, design is capable of drawing upon prototypes, experiments and mistakes, pilot projects, and speculation based on limited knowledge to sketch several paths that can describe the space for possibilities," writes the German graphic designer and university lecturer Florian Pfeffer.[32]

WHY IS IT SO HARD FOR US TO RELINQUISH CONTROL?

Ironically, there are designers who do exactly this while supported by robots and algorithms. Achim Menges is the director of the Institute for Computational Design (ICD) at the University of Stuttgart, where, after years of research conducted together with a large interdisciplinary team, he developed the *Elytra Filament Pavilion,* an extremely light, robot-constructed roof construction of carbon fibre and fibreglass which was first displayed at London's Victoria & Albert Museum in 2016

before being temporarily relocated to the Vitra campus in Weil am Rhein in February 2017. The individual modules are based upon biomimetic principles and are inspired by the wing cases of flying beetles known as "elytra". The modules themselves were designed by algorithms. Only a few individual parameters were predetermined, such as the fact that all of the modules should consist of hexagonal metal frames. However, the frames' exact geometry and structure of the fibres vary according to the bearing load, light, and weather conditions as well as the number of visitors. "In this instance the computer is more than just a tool," says Menges, "for it provides for levels of access that one otherwise would not have. One could compare it to a microscope or telescope, which do not change the world, but our view of it." He explains how computers are capable of dealing with complexities that go beyond the realm of human intuition. "This certainly does not mean that this is something that I would wish to simulate or control."[33] As a reward for this "controlled loss of control", the *Elytra Filament Pavilion* surprises designers with its unusual and fascinating aesthetics.

Achim Menges with Moritz Dörstelmann (ICD University of Stuttgart / Achim Menges Architect), Jan Knippers (ITKE University of Stuttgart / Knippers Helbig Advanced Engineering), and Thomas Auer (Transsolar Climate Engineering / TUM). *Elytra Filament Pavilion in the Victoria and Albert Museum,* 2016, photo: © NAARO, courtesy ICD, University of Stuttgart

15

30 "First Things First Manifesto 2000", in *Eye* (no. 33, vol. 8, Autumn 1999: no. 51, 1999), http://www.eyemagazine.com/feature/article/first-things-first-manifesto-2000, accessed on 4 December 2016.
32 Bruce Sterling, *Design Fiction,* http://shelovestofu.com/blog_uploads/2009/04/sterling-design-fiction.pdf, accessed on 4 December 2016.
33 Florian Pfeffer, *To Do: Die neue Rolle der Gestaltung in einer veränderten Welt – Strategien, Werkzeuge, Geschäftsmodelle* (Mainz, Hermann Schmidt, 2014), p. 176.
33 Achim Menges in a discussion with Amelie Klein (Weil am Rhein, 10 November 2016).

34 Chris Rehberger in a discussion with Amelie Klein (telephone interview, 26 October 2016).

The book which you are now holding in your hands was also designed by an algorithm devised by the Berlin graphic design firm Double Standards working together with a programmer. Here, too, a few fundamental parameters were determined – the basic raster, the fonts, a palette of type sizes, several options for illustrations, etc. – but the computer was given control over the overall design. At a touch of a button it came up with hundreds of thousands of layout options. Human designers were only responsible for selecting the final version, and, as Double Standards founder Chris Rehberger explains, they were better prepared to "try the impossible, for the algorithm schools the eye".[34] The result does not always fit with our reading and viewing habits, but we must remember that the graphic design and typography of the legendary Bauhaus books were also out of step with contemporary reading and viewing habits. After all, before Bauhaus there was no typesetting that allowed people to understand a text in its visual entirety at first glance.

The essence of experimentation is the process, not the goal. Perhaps the next book designed by Double Standards and its algorithm will do even more to shake up our habits – perhaps less. But for now this does not matter, just like it does not matter that the *3D Printed Cantilever Chair* designed by the CurVoxels student group takes itself to the point of absurdity. After all, it really isn't necessary to develop your own 3D printing software if you're setting out to produce the perfect *Panton Chair,* the design that served as a model for CurVoxels. The tried and true injection moulding process is certainly sufficient – after all, the *Panton* was designed to take advantage of the technique. The team's goal was not to print an improved version of the chair, however, for what they really wanted to do was experiment on an old complex form using complex new methods. A voxel is a three-dimensional pixel or, to explain it in different terms, a pixel in space. The *3D Printed Cantilever Chair* sets out to test – once again with the aid of an algorithm – just how many of these voxels a cantilever such as the *Panton Chair* requires in order to function properly. How much is possible with the minimum of material? A robot traces over the algorithmically determined path with hot plastic thread which solidifies while it is still in motion.

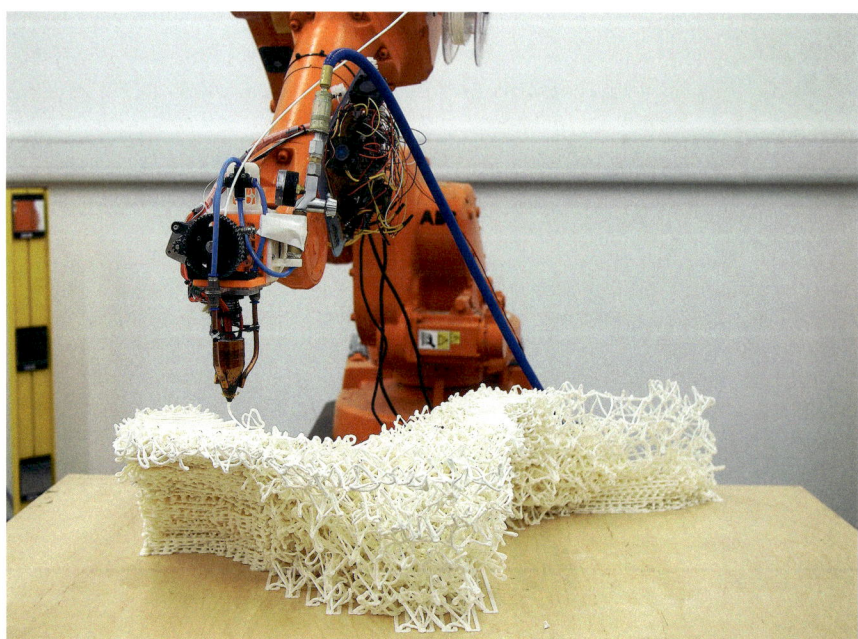

CurVoxels, Research Cluster 4, UCL The Bartlett School of Architecture, London. *3D Printed Cantilever Chair,* 2015. Chair and software for a 3D-printing technique. Team: (CurVoxels) Hyunchul Kwon, Amreen Kaleel, Xiaolin Li; Tutors: Gilles Retsin, Manuel Jiménez Garcia; Technical Support: Vicente Soler Senent, William Bondin © 2017 CurVoxels, photo: Sin Bozkurt, CurVoxels

Amelie Klein (born in 1971 in Vienna, Austria) is one of the curators of *Hello, Robot. Design between Human and Machine.* Since August 2011, she has been working as a curator at the Vitra Design Museum, most recently for the exhibition *Making Africa – A Continent of Contemporary Design,* for which she was nominated for the 2015 *ART* Magazine Curator Prize. Prior to this, Klein completed an MA in Design Criticism in New York and worked as Design and Creative Industry Editor at the Austrian daily *Die Presse.* She has published numerous articles in a range of design and architectural publications, including *Abitare, Domus Online,* and *Metropolis.*

William Williams. *The Cast Iron Bridge Near Coalbrookdale,* 1977. Oil on canvas, 86 × 102 cm. © courtesy Coalbrookdale Museum of Iron

For some time now, it has seemed as if we find ourselves at what could be described as the "Iron Bridge moment" of digital modernity. The Iron Bridge, built in the English county of Shropshire, is the world's first arch bridge to be constructed completely of cast iron. Yet even though it was built using what was then a fantastic new material, its construction rigidly follows the principles of wooden bridge design. It would take a few decades before the production and use of cast iron had been mastered to a degree that would eventually bring about a completely new aesthetic that was the natural result of the material's properties. Over the last few years we have certainly seen a number of 3D-printed "Iron Bridges", but the three examples described above provide us with a clue of the kind of aesthetics with which the early twenty-first-century will make it into the history books, if only we would learn to trust the algorithms and allow them to take control just for a moment. Perhaps we will one day come to accept that even though we may not be able to influence everything, something good can still result – such as an exciting roof construction, a new language of graphic design, or, to return to László Moholy-Nagy, life as a whole.

What does this all mean when it comes to how we deal with tyrannical forks and pill bottle tops that spy on us? Will it be enough if we – each and every one of us analogously to the scenarios outlined above – determine the parameters that can describe the scope of these smart devices and decide where humans take over again? Hardly. In this respect we are only now beginning to ask the right questions. You will find fourteen of them in this book and in the exhibition rooms of *Hello, Robot. Design between Human and Machine.* At first they might seem rather simple, but if you consider them more carefully you will soon realise that there are no simple answers. This, too, is a characteristic that weaves its way through the entire subject and reflects our postmodern world: there's no such thing as a single truth, for the contradictory strands of truth are often capable of existing alongside one another. But our fourteen questions invite visitors and readers to enter into a dialogue and reflect upon their own very personal relationship with technology as individuals but also as members of society as a whole.

But this is just the beginning. There is so much more to do.

SCIENCE AND
FICTION

Very few people have actually encountered a robot – at least one they would describe as such. This is because our ideas about and expectations of robots are strongly shaped by popular culture. From a young age we learn in films, TV series, books, comics, video games, and toys what robots look like, how they communicate with us, and how they behave: they are essentially like people, but made of metal. And what we all expect – more or less consciously – is that one day soon, we are either going to live with robots just as we do with our friends, neighbours, and colleagues or we are going to have to defend ourselves against them before they replace us once and for all.

Our fixation with humanoid robots ranges from the early fascination with automatons to the scientific laboratories of the present. But today's researchers really ought to know better, because, in reality, we are surrounded by robots and robotic systems that are capable of assuming every conceivable physical or digital form, materiality, scale, and intelligence level: from drones to self-checkout counters, from cranes to nanobots, and from vacuum cleaners with the intelligence of an amoeba to online chatbots that can engage us for hours. Cars and washing machines are partly robotic today and, ultimately, we can define any object and system as a robot if it can gather and store information from its surroundings, generate outputs that manifest themselves in some physical form, and display a degree of learning capacity and autonomy in the process.

The complex variety of robots' appearances is matched only by the complexity and ambi-valence of our relationships with them. The question of whether we need, or even like them is not really ours to ask; robots are already here, and as with smartphones, which most people didn't find necessary just a decade ago, there will one day be a critical mass of people who use smart, autonomous objects and applications and who will drag even the most ardent luddites willy-nilly into the robotic age. Whether robots will then be our friends or our enemies, or whether we will control them or vice versa, remains to be seen. And the question whether we should trust robots is perhaps less pertinent than whether we should trust the political-economic complex of humans, organisations, and infrastructure that stands behind them.

Our notion of what a robot is, what it looks like, and how it acts is inseparably linked to the utopian and dystopian visions of science fiction, as well as the pop culture images related to these visions. The fascination – and suspicion – of the machine that is potentially superior to humans characterises our primal fear of "superhuman" technology. The conceptual history of the term "robot" – derived from the Slavic (Czech) word *robota* meaning "forced labour" or "serf" – already predetermined our distrust of the powerful "artificial worker", which has long since become part of our society: through computer technology, automation, artificial intelligence, and algorithms.

**MARLIES
WIRTH**

THROUGH THE LOOK-ING GLASS, DOWN THE RABBIT HOLE: A MATTER OF TRUST

The term "robot" is documented for the first time in the theatre play *R.U.R. – Rossumovi Univerzální Roboti* ("Rossum's Universal Robots"), written by Karel Čapek in 1920.[1] The play refers to a species of robotic "workers" (from today's perspective more like androids or cyborgs), who serve humans until they eventually start a rebellion that results in the destruction of mankind. Just like the the Golem in Jewish Mysticism,[2] the creature in Mary Shelley's *Frankenstein or a Modern Prometheus* (1818), or the army of broomsticks that won't stop carrying buckets of water in Goethe's poem *The Sorcerer's Apprentice* (1797), countless examples from film and literature still reproduce the plotline of the "artificial human" who rebels against its creator: the robot Maria in *Metropolis* (1927), HAL 9000 from Stanley Kubrick's *2001: A Space Odyssey* (1968), and Ava from *Ex Machina* (2016) have more recently left their mark on the public's consciousness. All of these interpretations bring to mind the contemporary discourse on autonomous, self-learning machines and the dangerous potential of a future super-intelligence unbound.[3]

Nevertheless, the manual on how to use and benefit from the advantages of intelligent machines without having to fear them also derives from popular culture. The Russian-American biochemist and science fiction author Isaac Asimov sought to portray a more positive – and from today's perspective possibly more realistic – image of the "robot". In his 1942 short story "Runaround" Asimov framed the "Three Laws of Robotics" (also called "Asimov's Laws"),[4] which are based on three logical (hierarchical) principles: (1) A robot may not injure a human being or, through inaction, allow a human being to come to harm. (2) A robot must obey the orders given it by human beings except where such orders would conflict with the First Law. (3) A robot must protect its own existence as long as such protection does not conflict with the First or Second Laws.

With these laws Asimov shifted the focus to ethical issues, which are also relevant to science: Going beyond utopia and speculation, real technology, robots, and artificial intelligence have to be designed and implemented according to certain requirements that meet the ethical standards of humans. Considering the development of increasingly autonomous robots and self-learning artificial intelligence, not only do we need to thoroughly revise Asimov's Laws but also to appeal to the moral responsibility of the humans operating the technology, usually for-profit corporations and enterprises rather than individuals, or, in the best case scenario, largely independent research institutions.

1 Cf. Adam Roberts, *The History of Science Fiction* (New York, Palgrave Macmillan, 2006), p. 168.
2 The legend of the Golem, which has been passed on since the Middle Ages, describes how savants form a humanoid creature out of mud. The creature is able to perform tasks delegated by humans but can also destroy man due to its enormous size and power.
3 Cf. Nick Bostrom, *Superintelligence. Paths, Dangers, Strategies* (Oxford University Press, 2014), p. 28
4 Cf. Isaac Asimov, *I, Robot* (New York, Gnome Press, 1950).

FROM THE AUTOMATON TO AUTONOMY

While the term "automaton" (from *automatos,* Greek for "self-acting") describes a machine that is "capable of executing predetermined processes independently", the distinction and differentiation of the concept "robot" is much more difficult to grasp. Attempts to define it are usually based on the two terms "automatic" and "autonomous" (from *autonomos,* Greek for "acting by its own laws"). The robot can therefore be described as a *"sensor-motoric machine designed to expand human agency".*[5]

Considering the rapidly changing developments within the areas of robotics and artificial intelligence, five revised "robotic laws" were published by members of the Engineering and Physical Sciences Research Council (EPSRC) during a meeting of the European Association for Cognitive Systems (EUCog) in October 2013. Compared to Asimov's Laws, these new laws respond much more strongly to the actual status of robotic technologies and their coexistence with humans. At the same time, the laws link the idea that a robot is both a "tool" and a "product" designed to serve humans to the definition of the concept.[6]

The laws also stipulate ethical principles such as respecting existing fundamental rights and responsibilities – including freedom of expression and privacy – as well as that robots should not be used to kill (exceptions are made in the interests of "national security", which actually raises further ethical problems) or to "exploit" their users in any way (data collection, Big Data). Furthermore, the "machine-like nature" of the robot should remain transparent to humans at all times (an argument against the humanoid appearance of robots). The first and last principles emphasise human responsibility for robots, including de facto legal responsibility and fundamental ethical responsibility.

FROM "OPERATING" TO AGENCY

In a society of human and non-human beings, all natural, social, and technical "objects" are regarded not only as solely constituted by society but also as co-constituting society. In any case, the distinction between them is less relevant than the question of the capabilities and dangers of their agency and, moreover, to what extent their actions – in contrast to automatic operation and execution – influence the hierarchical structure of a society.[7]

Marshall McLuhan's media theory[8] ("The Medium Is the Message") makes clear that not only the content of a "medium" but also its characteristics influence the society to which it is relevant. While a "thing" is constituted as an object of utility which is able to fulfil a particular function, a medium assumes a role in which it communicates or transmits; therefore, it is not simply itself.[9] This also corresponds, for instance, to the "actor-network theory" (ANT) developed by Bruno Latour, according to which acting is not limited to humans ("actors"), but extends to non-human entities ("actants").[10]

5 Definition by Thomas Christaller in: Thomas Christaller, et. al., *Robotik. Perspektiven für menschliches Handeln in der zukünftigen Gesellschaft* (Berlin, Springer, 2001).
6 Original text see: https://www.epsrc.ac.uk/research/ourportfolio/themes/engineering/activities/principlesofrobotics/
7 Cf. Bruno Latour, "Where Are the Missing Masses? The Sociology of a Few Mundane Artifacts", in Wiebe E. Bijker and John Law (eds.), *Shaping Technology / Building Society: Studies in Sociotechnical Change* (Cambridge, MA, MIT Press, 1992), pp. 225–258.
8 Marshall McLuhan, *Understanding Media: The Extensions of Man* (London, Routledge, 1964/2005).
9 Mercedes Bunz, "Die Dinge tragen keine Schuld. Technische Handlungsmacht und das Internet der Dinge", in Florian Sprenger, Christoph Engermann (eds.), *Internet der Dinge. Über smarte Objekte, intelligente Umgebungen und die technische Durchdringung der Welt* (Bielefeld, Transcript Verlag, 2015), p. 169; Cf. Fritz Heider, *Ding und Medium* (Berlin, Weltkreis-Verlag, 1927).
10 Bruno Latour, "On Actor-Network Theory: A Few Clarifications", in *Soziale Welt* (no. 47, 1996, book 4) pp. 369–382. http://www.bruno-latour.fr/sites/default/files/P-67%20ACTOR-NETWORK.pdf

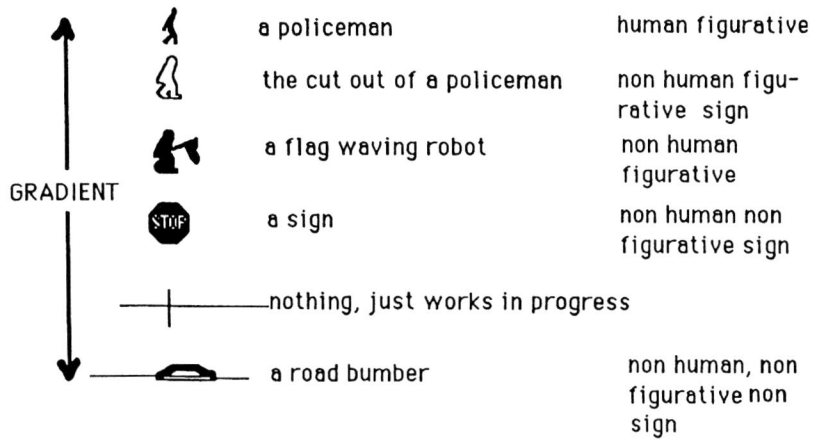

a policeman	human figurative	
the cut out of a policeman	non human figurative sign	
a flag waving robot	non human figurative	
a sign	non human non figurative sign	
nothing, just works in progress		
a road bumber	non human, non figurative non sign	

GRADIENT

Bruno Latour in *Where Are the Missing Artifacts;* with the kind permission of Burno Latour, courtesy The MIT Press, in *Shaping Technology / Building Society: Studies in Technological Change,* Wiebe Bijker and John Law (eds.)

In this context, the definition of "action" is under consideration, too. In her magnum opus, *The Human Condition* (1958), the philosopher Hannah Arendt uses three terms to define the basic conditions of human life that also describe the individual's autonomous, active participation in society: labour, work, and action. While Arendt's understanding of labour and work subsumes those (individual) activities which are immediately necessary for the production of (material) goods, she describes (inter)action – language and communication – as humans' greatest asset. After the three areas identified by Arendt are increasingly taken over by robots, AI, and algorithms, Bruno Latour's theory becomes applicable insofar as these autonomous entities do not just merely transfer action, they also perform it. Thus, as autonomous entities, they don't merely participate in the society of humans but also actively (co-)construct it.

Nevertheless, the robot's potential ability to act doesn't relieve humans of their responsibility; by equating human with machine we eventually risk letting both the intention of the action and any responsibility and accountability become neglected.[11]

11 Felix Stalder, "Beyond Constructivism: towards a Realistic Realism: A Review of Bruno Latour's *Pandora's Hope,*" in *The Information Society* (vol. 16, no. 3, 2000), p. 4; http://felix.openflows.com/html/pandora.html

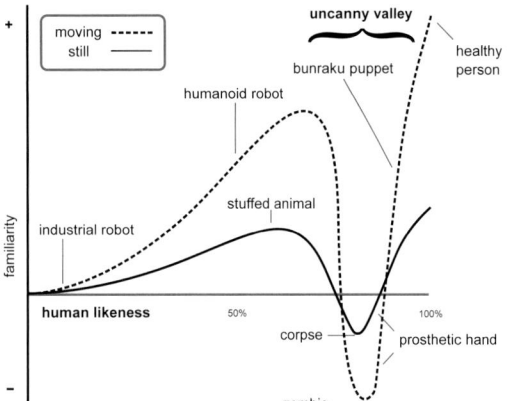

Masahiro Mori's "Uncanny Valley" defines the amplifications in the emotional response of humans in relation to the degree of anthropomorphism of a robot. "Uncanny Valley" denotes the zone in which people show a negative reaction to robots that appear "too human". © Smurrayinchester, self made graphic based on an image by Masahiro Mori and Karl MacDorman from http://www.android-science.com/theuncannyvalley/proceedings2005/uncannyvalley.html CC BY-SA 3.0

24

12 The first autonomous "Soft Robot" in the shape of an octopus was developed by scientists at Harvard University in August 2016. Cf. http://news.harvard.edu/gazette/story/2016/08/the-first-autonomous-entirely-soft-robot/
13 Cf. Masahiro Mori, "Uncanny Valley", translated by Karl F. Mac-Dorman and Norri Kageki, 2012: http://spectrum.ieee.org/automaton/robotics/humanoids/the-uncanny-valley

HUMAN VS. HUMANOID

Although both popular culture and science are obsessively committed to the anthropomorphic appearance of robots,[12] it has been proven that this significantly "human" – and hence confidence-inspiring – aspect isn't just transmitted through visual perception; it also depends on many additional factors such as smell, body expressions and gestures, haptics or vocabulary. Consequently, the feeling of "familiarity" decisively influences the acceptance of a robotic counterpart.

Deceptive human resemblance can even have the opposite effect: With his concept of the "uncanny valley", the Japanese roboticist Masahiro Mori defined in 1970 the eerie feeling that overcomes humans when they are dealing with a human-looking but artificial "other". The "uncanny valley"[13] describes the rapid drop in acceptance observed when people are presented with objects displaying a very high degree of anthropomorphism. On the other hand, abstract or abstracted figures and recognisably artificial entities are more accepted, as they do not conceal their artificiality. The "uncanny" occurs in the valley-shaped zone of this acceptance curve.

14 Andrea Sick, "Questions of Style: Subjects, Things and Shared Agency in Popular Articulations", in Christoph Lischka, Andrea Sick (eds.), *Machines as Agency: Artistic Perspectives* (Bielefeld, Transcript Verlag, 2007), pp. 122, 123.
15 Ibid., p. 123.
16 Ibid., p. 123; cf. Sibylle Krämer, "Maschinenwesen. Ein Versuch über den Anthropomorphismus in der Technikdeutung hinauszukommen", in Thomas Christaller, Josef Wehner, *Autonome Maschinen* (Wiesbaden, Westdeutscher Verlag, 2003), pp. 208–221.
17 Alan Turing, *Computing Machinery and Intelligence,* (Aberdeen, University Press, 1950).

With the growing interaction between human and machine, the question of trusting non-human others is linked to the role and function of a robot's "similarity to humans", which goes beyond its humanoid appearance. In *Machines as Agency,* Andrea Sick proposes three fundamental versions as possible explanations:[14] Assuming that there is an ontological difference between humans and technology, which is based on the attributions "natural" and "artificial", the anthropomorphism of the machine is not so much a question of its form or appearance, but rather one of its *genesis* (origin). Following the theory of "human enhancement" and the expansion of our possibilities through technology (cf. McLuhan: "media as the extensions of man"), humans and machines basically have the same function, yet machines serve humans to reach a higher functionality. Hence, the machine could adopt any conceivable or necessary shape. The third possibility states the use of technologies as a paradigm for explaining human skills: "The technical emerges as a medium in the process of modelling and discovery."[15]

All three possibilities assume a functional correspondence between human and machine. Thus, the machine must be able to deceive or pretend to be potentially human. The ability to deceive humans, that is, to simulate humanity and human[16] "consciousness", is described as a success in the "Turing Test": The British logician and computer scientist Alan Turing developed the question-answer-interaction – which is still used today – with two conversation partners, one human and one artificial intelligence. An artificial intelligence passes the test if the questioner is not able to distinguish between human and machine.[17]

TRUST: "MACHINES WILL MAKE BETTER CHOICES THAN HUMANS"

Trust is not a question of technology but rather a social decision.[18] Whenever we delegate tasks to machines and robots, we automatically trust them. However, by "trusting" robots we assume that their foundations for decision-making are built on the same level of understanding of social and ethical principles as humans'.[19] Confidence can be described as a state that arises in a network of relationships between human beings; this state, in turn, is what enables humans to develop such a feeling at all. Trust is located in the grey area between knowledge and ignorance and, similar to doubt, describes a feeling that enables action despite a lack of certainty.[20] We attribute responsibility to the person we trust and if they misuse that trust we can "hold them responsible" for the abuse of our confidence.[21] But how can we extend this argument to a robot or a "non-human actant"?

A study conducted by MIT – Massachusetts Institute of Technology – invites visitors to a website named "Moral Machine" to test which moral decisions a self-driving car (or for now its human programmer) has to be able to make.[22] The ethical dilemma, which occurs in case of an unavoidable accident, is whether the occupants of the vehicle or children, elderly or socially disadvantaged people should be treated preferentially. Like in the thought experiment described as the "trolley problem",[23] a single "right" decision is impossible. So how can we trust autonomous vehicles?

18 Bruno Latour, *Die Hoffnung der Pandora. Untersuchungen zur Wirklichkeit der Wissenschaft* (Frankfurt am Main, Suhrkamp Verlag, 2000; 5th ed, 2015), p. 230 ff.
19 Mark Coeckelbergh, Can we trust robots?, Springerlink.com, 2011, p. 1; http://link.springer.com/article/10.1007/s10676-011-9279-1
20 Vilém Flusser, *On Doubt* [1966] (Minneapolis, Univocal, 2014), pp. 3–4.
21 Mark Coeckelbergh, see note 19, p. 3.
22 http://moralmachine.mit.edu/
23 Philippa Foot, "The Problem of Abortion and the Doctrine of the Double Effect," in *Virtues and Vices* (Oxford, Basil Blackwell, 1978; originally published in Oxford Review, no. 5, 1967).

24 http://spectrum.ieee.org/autom-aton/robotics/artificial-intelligence/researchers-teaching-robots-how-to-best-reject-orders-from-humans

25 Microsoft developed the Twitter-bot "Tay", which had to be taken off the system after only 24 hours of use, because it had learned and adopted racist and unethical state-ments by Twitter users; Cf. http://www.theverge.com/2016/3/24/11297050/tay-microsoft-chatbot-racist An algorithm used as a judge in a beauty contest preferred women with light skin to those with darker skin; Cf. https://www.theguardian.com/technology/2016/sep/08/artificial-intelligence-beauty-contest-doesnt-like-black-people

With this context in mind, in 2015 scientists at the Tufts University Human-Robot Interaction Lab (Boston, Massachusetts) experimented with mechanisms that allowed robots not to obey human commands under certain conditions.[24] The experiment mainly referred to a refusal of the chain of command based on ambiguous language (if the command is not understood, it can be rejected) or a sensomotoric evaluation by the robot (if there is no solid ground, moving forward can be denied). Still, the experiment negates Asimov's Second Law in a sense ("A robot must obey the orders given it by human beings...") and thereby fundamentally challenges the logic of Asimov's hierarchy, even though the action did not result in danger or injury to a human.

"Wrong" decisions made by robots and artificial intelligence today are merely based on the fact that machine learning data is still provided by humans and thus, existing views and errors are passed on and continued.[25] But what happens if artificial intelligence develops so far beyond human control that we can no longer regulate what they learn and which action they take as a result?

26 Original text of the open letter "Research Priorities for Robust and Beneficial Artificial Intelligence" on the website of the Future of Life Institute: http://futureoflife.org/ai-open-letter/

27 Cf. Nick Bostrom, see note 3.

28 Interview with Nick Bostrom in *The Guardian,* June 2016: https://www.theguardian.com/technology/2016/jun/12/nick-bostrom-artificial-intelligence-machine

29 Ray Kurzweil, *The Singularity Is Near: When Humans Transcend Biology* (New York, Viking Books, 2006).

In January 2015, leading scientists including Nick Bostrom, Professor of Philosophy and Director of the Future of Humanity Institute at Oxford University, physicist and astrophysicist Stephen Hawking, and Elon Musk, founder of SpaceX and co-founder of Tesla Motors, wrote an open letter in which they warned of an "Intelligence Explosion".[26] Numerous internationally renowned experts also signed the letter, in which the scientists argue in favour of creating priorities in the field of AI research that foster a "robust and beneficial artificial intelligence". Their fear was a "super-intelligence",[27] resulting from the rapid development of machine learning and deep learning, could lead to mankind no longer being able to evaluate potential consequences and to eventually being overtaken by technology.[28] The futurologist Ray Kurzweil even predicted a new "collective consciousness" – the "singularity" – an evolutionary process over the course of which human and machine would continuously assimilate until they became one.[29]

The fact that hyper-intelligent robots with the ability to self-learn and autonomously reproduce and / or independently repair individual parts are mostly being developed in the context of military technology may leave an unpleasant aftertaste. Trusted networks, data encryption, decoding software, authentication, and all other infrastructures relevant to communication have been created for and are controlled by the military. Whereas humans are becoming more and more "transparent", technology is turning into an inscrutable "black box".

Communicating objects that study our habits and send all relevant data into the "cloud" surround us on a daily basis. It is impossible to trust that these data are not accessible to outsiders (people, corporations, the state, or terrorists). The notion of confidence and, hence, the guarantee of human privacy, implies a system that is not capitalistically corrupted (decentralised instead of monopolised) – hence, humanity needs a comprehensive overhaul of its political structures and to establish a new cultural consciousness. The call to trust in technology is merely a matter of trust in humans.

Marlies Wirth (born in 1980 in Neunkirchen, Lower Austria) is one of the curators of *Hello, Robot. Design between Human and Machine*. She was recently named Curator Digital Culture & Design Collection at the MAK Vienna, where she has worked since 2006. She has curated exhibitions, performances, and discursive events in the fields of art, design and architecture including the *Hollein* retrospective (2014) and the themed group show *24/7: the human condition* for the Vienna Biennale 2015. With a focus on conceptual, site-specific, research-, and time-based art and a particular interest in the cultural-anthropological contexts of artistic production, she also develops independent exhibition projects with international artists.

Considering the discourse on robotics, artificial intelligence, and the far-reaching effects of automation and digitalisation in all areas of human existence, it is ultimately a discourse about humans and the implications of these developments that inevitably affect our quality of life, our individuality, our decisions, our creativity, our mental and physical health, and our human society in a global sense. With regard to the political and social tendencies within the last decade, there is one burning question still to be answered: What defines "humanness" in the twenty-first century? And how do we deal with the urgent problems and conflicts of our time that cannot and will not be solved by robots and algorithms? Which ethical, philosophical, and socio-cultural principles provide the foundations for present and future technological progress that will enable us to act in the interests of continued human existence? And last but not least, which questions must be asked in order to make these decisions at all?

There is a defining ambivalence in our minds when faced with the topics of the digital age that results in uncertainty about what is "right" or "wrong", "good", "bad", or "better". This ambivalence is as omnipresent as it is relevant, especially in the context of an increasingly complex world which is not solely determined by humans anymore. Unlike binary-based computer logic, humans have the ability to act without certainty: to doubt and to trust at the same time. Spanning science and fiction, *Hello, Robot. Design between Human and Machine* shows the – often invisible – design elements that define the coexistence of human and machine, and tells of work and play, of identity, loneliness, violence, love, fear, and trust.

DO YOU
TRUST
ROBOTS?

DO
NEI

HAVE YOU EVER M
ROBOT?

VE REALLY
) ROBOTS?

T A

WHAT
WAS
YOUR
FIRST
EXPERI-
ENCE
WITH A
ROBOT?

RE ROBOTS OUR
RIENDS OR OUR ENE-
IES?

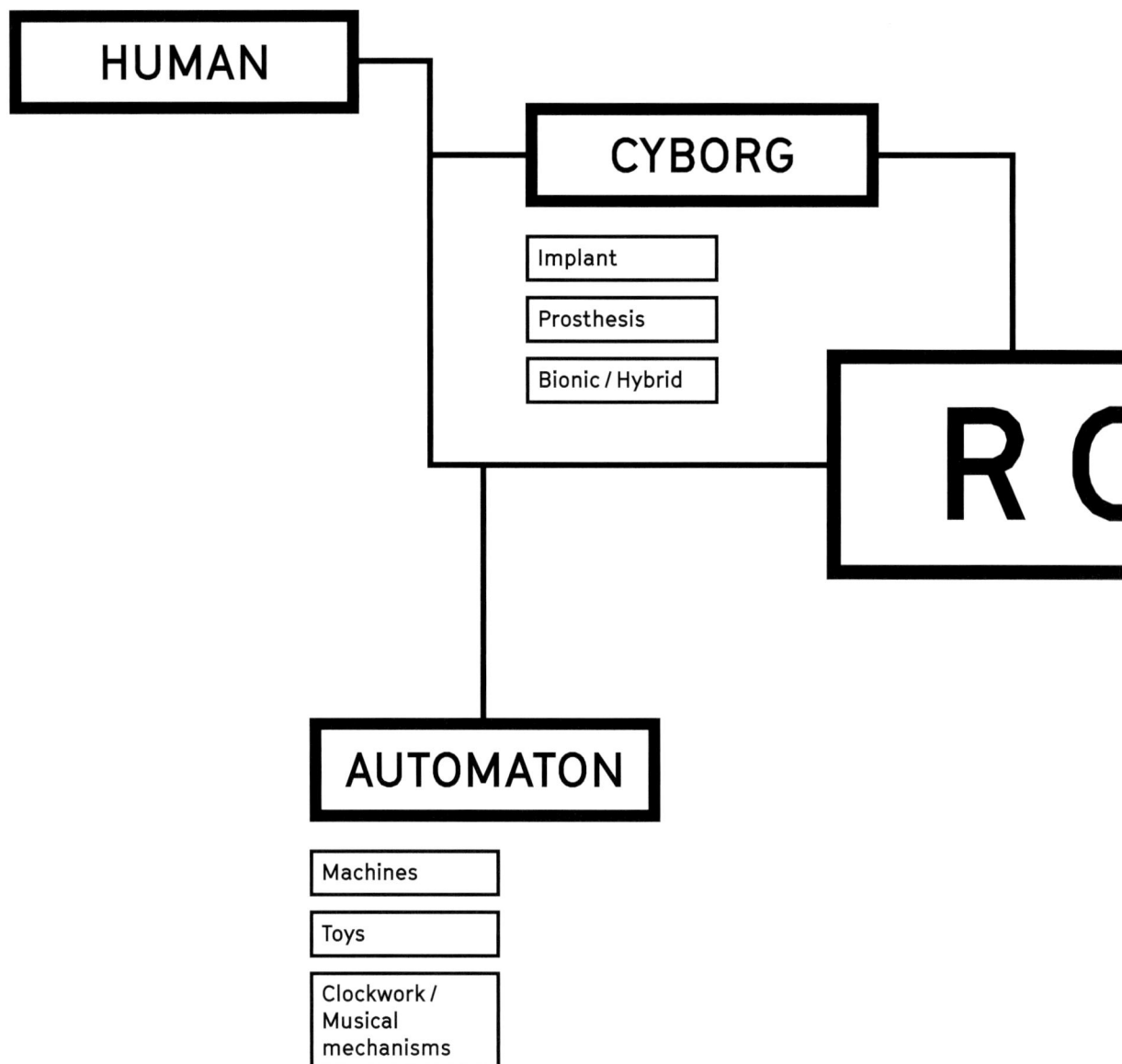

HUMAN

CYBORG

Implant

Prosthesis

Bionic / Hybrid

R C

AUTOMATON

Machines

Toys

Clockwork /
Musical
mechanisms

THE ROBOT
APPROACHING A SPECIES

SOFTWARE

BOT

ARTIFICIAL INTELLIGENCE (AI)

DISTRIBUTED INTELLIGENCE

AUGMENTED / VIRTUAL
REALITY (AR / VR)

HARDWARE

HUMANOID	CREATUROID	MACHINE	INFRASTRUCTURE / UBIQUITY
Anthropomorphic	Animal-like	Industry / Digital Production	Smart Devices
Woman-like (gynoid)	Alien-like	Science / Education	Internet of Things
Man-like (android)	Robot-like	Military	Smart Home
		Transportation / Logistics	Smart City
		Sustainability / Environment	Sensory Environment
		Medicine / Care	
		Art / Entertainment	
		Household	

@ Paul Feigelfeld, Vitra Design Museum,
Graphic: Hug & Eberlein, Leipzig—Basel

ROBERT R. SNODY – *THE MIDDLETON FAMILY AT THE NEW YORK WORLD'S FAIR*

The Middleton Family at the New York World's Fair, 1939

Robert R. Snody. *The Middleton Family at the New York World's Fair,* 1939. Film, 55 min © 2016 courtesy the Internet Archive

The Middleton Family at the New York World's Fair is an American film produced by Westinghouse Electric Company for its exhibition at the 1939 New York World's Fair. The film was designed to represent the middle-class response to domestic products and technology-driven social improvements in an imagined future that was presented at the fair and that included dishwashers and demonstrations of robots. "Elektro, the Moto-Man", was on show at the fair in 1939 and reappeared in 1940. It was a 2.1m tall, 120.2 kg robot that could walk, respond to voice commands, speak about 700 words, move its head and arms, distinguish between colours, blow up balloons, and even smoke cigarettes. In 2012, the film was one of the 25 motion pictures added that year to the United States National Film Registry. AR

Gundam is a Japanese science fiction media franchise created by Sunrise. The franchise started in 1979 as an anime TV series called *Mobile Suit* Gundam, and today comprises movies, manga, novels, video games, and toys. *Gundam* defined the "real robot" genre in anime by featuring giant robots (or "mecha") in a militaristic war setting.

Most Gundam are large, bipedal, humanoid-shaped vehicles controlled from cockpits by a human pilot. The "mobile suits" have a cockpit in the machine's "torso", while a built-in camera in the "head" transmits images to the cockpit. Gundam are non-sentient machines.

The *Zaku* was one of the first mobile suits units created for battle scenes in the series, and it is one of the most recognizable "mecha" designs in anime. AR

BANDAI
– *GUNDAM*
AND *ZAKU*

Gundam, 1995

Bandai Co., Ltd. *Gundam* and *Zaku,* 1995. Mobile Gundam suits, various materials, 150 × 80 × 40 cm each © Bandai Co., Ltd., Japan, photo: Vitra Design Museum

Zaku, 1995

Originally created for the 2015 interactive robot festival "Bal Robotov" in Moscow, *Robin*–whose sweet, round face fulfills our cliché of a friendly robot – has been labelled "transgender" by her creator, the Belgian robot designer Jan De Coster. When referencing *Robin,* De Coster uses female pronouns in order to simplify things, although he doesn't make clear whether "she" is in fact a she or a he. By doing so, De Coster questions the public's tendency to assign gender to robots, even ones that don't have faces or names as cute as *Robin's*. With a matte black face and hot air tube arms, "she" certainly doesn't fit the prevailing notions of femininity; but why are we even projecting? *Robin* is just a robot. EP

Jan De Coster. *Robin,* 2015. Interactive object, found materials, aluminium, plexiglass, hot air tubes,
3D printed parts, 56 × 46 × 46 cm
© Jan De Coster

JAN DE COSTER – *ROBIN*

Robin, 2015

NINTENDO – *R.O.B. (ROBOTIC OPERATING BUDDY)*

Nintendo. *R.O.B. (Robotic Operating Buddy),* 1985. Video game accessory, various materials, 24 × 16 × 16 cm © Nintendo

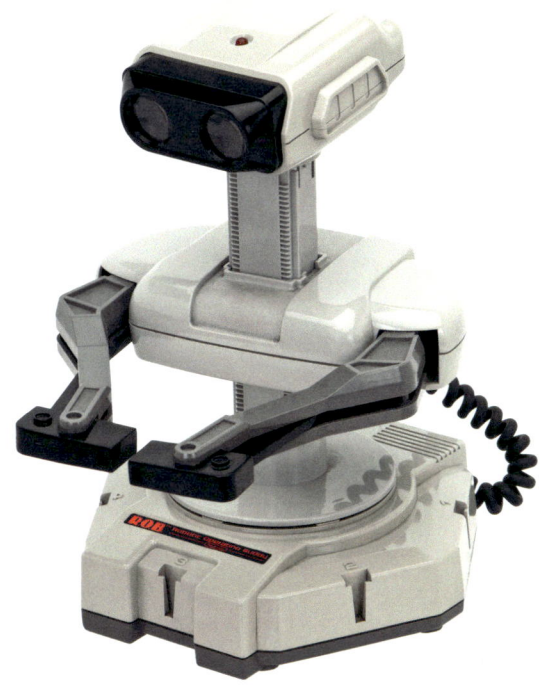

R.O.B., 1985

Nintendo's *Robotic Operating Buddy,* aka *R.O.B.,* was launched in 1985 and represents an early attempt to transport a virtual video game into off-screen reality. The player directs the robot, which receives commands via the screen, and causes it to complete simple routines, such as sorting rings of various colours in prescribed sequences.

At the time a minor attraction, *R.O.B.'s* level of technical sophistication meant that the interactions between human and robot were so slow and halting – especially by today's standards – that it was difficult to develop a flow while playing the game. Thus, over the long term, *R.O.B.* was unable to gain a following among Nintendo fans. LH

In 2016 students from the Bachelor's Programme Media & Interaction Design at the Swiss design school ECAL were commissioned to create an installation for the exhibition *Hello, Robot.*, which explores peoples' ambivalent relationship to robots. A one-week workshop led by Cyril Diagne (Dept. Head) and Alain Bellet gave students an opportunity not only to test the abilities of the Thymio robots, but also to question their ethical implications. The final installation, instigated by students Luca Kasper and Mathieu Palauqui, presents Thymios moving up and down wooden see-saw planks, creating a sense of measured uncertainty. The robots' repetitive movements allow them to spell out the sentence "error is human and to blame it on a robot is even more so", based on the well-known quote by the American comedy writer Robert Orben. EP

ECAL University of Art and Design, Lausanne, BA Media & Interaction Design. *Error Is Human,* 2016. Mixed media installation with Thymio robots (EPFL / ECAL / ETHZ / Mobsya); ECAL/students: Luca Kasper, Mathieu Palauqui; tutors: Alain Bellet, Cyril Diagne; assistant: Laura Perrenoud © 2016 ECAL

ECAL – *ERROR IS HUMAN*

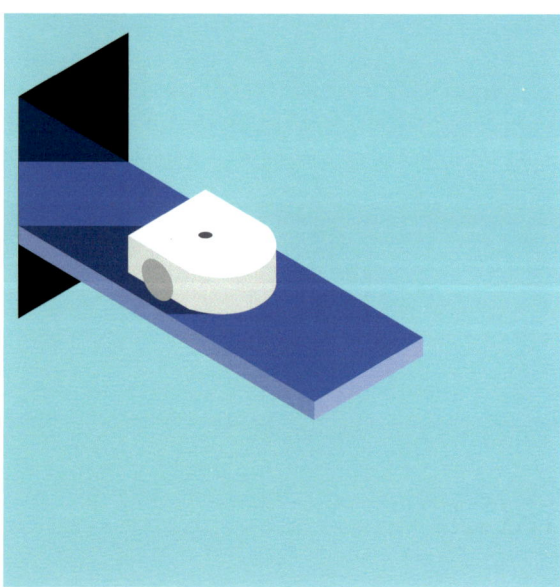

Error Is Human, 2016

In today's world, robots are already capable of relatively simple functions that allow them to perform everyday household tasks. Vacuuming robots, such as *Roomba* sold by the company iRobot, can be found in millions of homes across the globe. The robot automatically cleans floors provided they are on a continuous plane, and it has built-in sensors and a navigational system to tell it where it has been and where it still needs to clean. Even more interesting is the relationship that many owners have to their *Roombas*. A number of studies have shown that people often give them names and treat them as family members, just like they would their pets. TT

Roomba 980, 2016

IROBOT – *ROOMBA 980*

WHAT WAS YOUR FIRST EXPERIENCE WITH A ROBOT?

ARE ROBOTS OUR FRIENDS OR OUR ENEMIES?

AVE YOU EVER MET A OBOT?

DO WE REALLY NEED ROBOTS?

DO YOU TRUST ROBOTS?

At the centre of *Rossum's Universal Robots,* Karel Čapek's play from 1920, is the eponymous company R.U.R., which produces human-like clones out of synthetic organic materials. The clones serve as cheap labourers with no rights whatsoever – until they rise up in rebellion against their human masters. In Čapek's play – which was translated into a number of languages – the word "robot" makes its first appearance. It would soon become a common term in many countries around the world. The word's origins lie in the Czech word *robota,* which can be translated as "forced labour". *R.U.R.* also takes up the motif of the Golem, whose origins can be found in Jewish mysticism, and alludes to a leitmotif in the cultural history of the robot: the artificial being created for a specific purpose frees itself from human control and, in turn, poses a threat to humanity. TT

KAREL ČAPEK – *R.U.R. [ROSSUM'S UNIVERSAL ROBOTS]*

Vandamm Studio. Scene from *R.U.R,* 1928–29. Photograph of the New York stage production of the Karel Čapek play, Guild Tour Company, 20.3 × 25.4 cm © Billy Rose Theatre Division, the New York Public Library for the Performing Arts

Programme for *R.U.R. (Rossum's Universal Robots),* Frazee Theatre, New York, 1922. Programme, colour print on paper, 19 × 13.8 cm; Set design: Lee Simonson © Austrian Theater Museum

Playbill for *W.U.R. (Werstands Universal Robots),* Neue Wiener Bühne, Vienna, 1923. Playbill, print on paper, 25.5 × 20 cm © Austrian Theater Museum

A scene from the New York production of *R.U.R. [Rossum's Universal Robots],* 1928–29

Frederick Kiesler's set design for the 1923 production of Karel Čapek's 1920 play about robots, which originally went by the title *Werstands Universal Robots* in Germany, was a smash hit among Berlin's avant-garde. Kiesler's "electro-mechanical scenery" was aimed at enlivening the set design by incorporating technical and faux-technical means, such as an "interplay of coloured lights and floodlights" coordinated with the actors' speech and movements. TT

Frederick Kiesler. Electromechanic scenery for Karel Čapeks *W.U.R [Werstands Universal Robots]*, Theater am Kurfürstendamm, Berlin 1923. Photograph, gelatin silver print on baryta paper, 20.2 × 25.3 cm © 2017 Austrian Frederick and Lillian Kiesler Private Foundation, Vienna

Set design for the Berlin production of *W.U.R. [Werstands Universal Robots]*, 1923

FREDERICK KIESLER – SET DESIGN FOR *W.U.R. [WERSTANDS UNIVERSAL ROBOTS]*

Robots of Brixton, 2011: dystopian architecture, ...

... workers, ...

Kibwe Tavares. *Robots of Brixton,*
2011. Film, 6 min © Factory Fifteen

KIBWE TAVARES – *ROBOTS OF BRIXTON*

... and riots in the streets.

Robots of Brixton is a short architectural sci-fi film made by architect and director Kibwe Tavares, who was born and raised in Brixton. The film shows a bleak future of the district. London's vast population of robotic workers is cooped up in this area, resulting in unplanned, cheap, and hasty additions to the skyline. The robots, which constitute an oppressed workforce designed to carry out all the tasks that humans are no longer willing to do, end up battling with the police against a backdrop of dystopian architecture. The scenes showing this conflict echo the real-life Brixton riots of 1981. The social and architectural aspects of the movie earned several accolades for Tavares, including the Jury Prize for animation direction at the Sundance Film Festival and the RIBA Silver Medal. AR

44

Book cover design for *Das deutsche Wirtschaftswunder,* 1927

John Heartfield. Book cover design for *Das deutsche Wirtschaftswunder* (author: Günther Reimann), 1927. Book, 23 × 15.7 cm; source of original: Akademie der Künste, Berlin, Art Collection, Inv. No.: Heartfield 2336 © The Heartfield Community of Heirs / VG Bild-Kunst, Bonn 2016

JOHN HEARTFIELD – BOOK COVER DESIGN FOR *DAS DEUTSCHE WIRTSCHAFTSWUNDER*

In the wake of the technological revolution, the highly political Berlin branch of the Dada movement responded to the era's exploitation of workers with nonsense and satire. The economic policies of the day were fiercely mocked by the Dada artist John Heartfield's collages, which also feature his frequently used motif of the spectre. This figure appeared on the cover of Günther Reimann's 1927 book *Das deutsche Wirtschaftswunder,* which criticised new economic policies and worker exploitation. In lieu of a real body and soul, the factory worker has a stopwatch for a head and a time clock for a body. Exemplifying the "Maschinenmensch", the transformation of man into machine, this figure captured the zeitgeist of the era and influenced later depictions of the robot, a concept previously unknown. EP

Metropolis is a 1927 German Expressionist science fiction film directed by Fritz Lang. Set in a futuristic urban dystopia, *Metropolis* follows the attempts of the wealthy son of the city's ruler and a poor worker to overcome the gulf separating the classes of their city. In the futuristic city of *Metropolis,* automation has created rather than alleviated the drudgery of the workers. The film is considered a pioneering work in the science fiction genre and includes one of the first robots ever depicted in cinema. OP

Metropolis, film poster, 1926

FRITZ LANG – *METROPOLIS* (FILM POSTER BY HEINZ SCHULZ-NEUDAMM)

Heinz Schulz-Neudamm. *Metropolis,* 1926. Film poster, multicolour print, 211 × 96 cm © courtesy Austrian National Library, Vienna, Picture Archives and Graphics Collection

Isaac Asimov. "Runaround", in
Astounding Science Fiction, Vol.
XXIX, No. 1 (March 1942). Peri-
odical, cover art: Reginald Hubert
Rogers © 2016 by Penny Publications / Dell Magazines (Analogsf.
com), reprinted with the permission
of the publisher, photo: courtesy
University of Regensburg Library

ISAAC ASIMOV – "RUNAROUND", *ASTOUNDING SCIENCE FICTION,* 1942

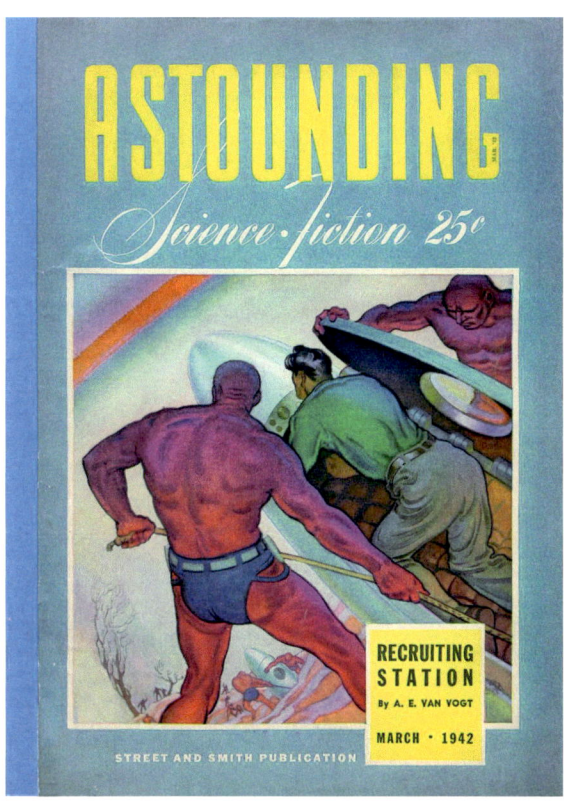

Astounding Science Fiction, magazine cover, 1942

Isaac Asimov was one of the most prolific authors of
all time, writing or editing more than 500 books
in his lifetime, all the while catering to a large general
audience. In 1939 he began publishing sci-fi short
stories in the magazine *Astounding Science Fiction*
(ASF), including "Runaround", which appeared in
1942. The story saw the first ever mention of Asimov's
"Three Laws of Robotics", which still serve as the
groundwork for the discussion on robots today, as
well as the first ever use of the word "robotics".
"Runaround" is the story of a robot that behaves in
an unusual and counter-intuitive way because it
takes the "Three Laws" literally.

The Three Laws are:
1) A robot may not injure a human being or, through
inaction, allow a human being to come to harm.
2) A robot must obey the orders given it by human
beings except where such orders would conflict with
the First Law.
3) A robot must protect its own existence as long as
such protection does not conflict with the First or
Second Laws.

Published since 1930, *Astounding Science Fiction,*
currently titled *Analog Science Fiction and Fact,* is
the longest-running continuously published science
fiction magazine in the history of the genre. AR

47

Astro Boy relates the adventures of an android boy equipped with superhuman powers called *Tetsuwan Atomu* in Japanese, which translates roughly as "Mighty Atom". He was created by a scientist as a substitute son after his own son had died. *The Astro Boy* mangas began appearing in 1951 and achieved world-wide fame after an English translation was published in the 1980s. The TV adaptation of 1963 was the first successful animated Japanese television series in the new style, which later became popular all over the world under the name *anime*. At the time, 40 percent of all Japanese people who owned a television set regularly watched the series. LH

Osamu Tezuka. *Astro Boy* (original Japanese title: *Tetsuwan Atomu*), 1952–1968. Science fiction manga, photo: October 1966 edition of Shōnen (Publ. Kobunsha) © 2017 Tezuka Productions

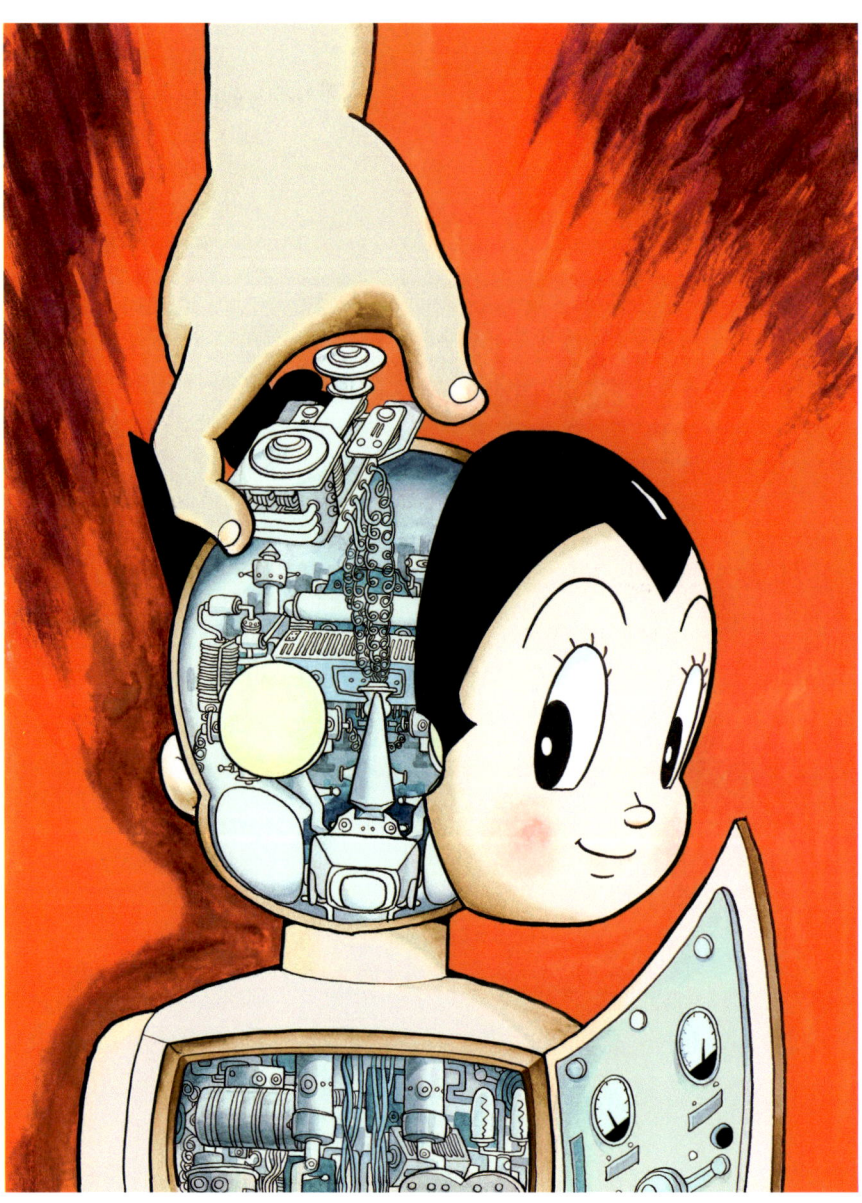

Astro Boy, 1966

OSAMU TEZUKA – *ASTRO BOY*

MATT GROENING AND DAVID X. COHEN
– *FUTURAMA*

49

Futurama, 1999

This American animated science fiction comedy series by Matt
Groening and David X. Cohen ran from 1999 to 2013. It tells
the story of a spaceship crew working for an interplanetary delivery service in the thirty-first century. Alongside Fry, a typical
antihero who hails from the twenty-first century, and Leela,
the spaceship's ambitious pilot, one of the stars of the series
is the robot crew member Bender, who maintains an eccentric
lifestyle full of smoking, cursing, and excessive alcohol consumption. A few episodes contain references to the robot HAL
from Stanley Kubrick's *2001: A Space Odyssey* (see p. 61), such
as the "HAL Institute for Criminally Insane Robots", a closed
asylum for psychotic robots. LH

VINTAGE TOY ROBOTS

Thanks to cheap metal and established tin printing methods, Japan rose to the top of the tin toy-making industry of the 1950s and 1960s with its unique toy robots. Children across the globe, especially in Cold War-era USA, were fascinated with the novel toys, which were released while the international race to put a man on the moon was in full swing. For the Japanese, the robots were meant to be seen as peaceful helpers; they were not equipped with weapons until spacemen with "Western" physiognomies were put into production. Owing to the fragility of their metal and its replacement by plastic, few of these original robots have survived, making them highly collectable objects today. EP

Various manufacturers. Vintage toy robots, 1956–ca. 1980. Various materials © Private collection, photo: Andreas Sütterlin

Vintage toy robots, 1956–ca. 1980.

... which is easy to understand ...

Robo Wunderkind, 2015: A modular robotic system, ...

ROBO TECHNOLOGIES – *ROBO WUNDERKIND*

Robo Wunderkind is a modular, colourful robot kit that allows children as young as five years old to have fun building their own robots and learning basic coding skills. The kit consists of smart cubes, each roughly 8 cm in size and containing a variety of embedded electronics such as cameras, proximity sensors, and laser pointers. They are connected wirelessly and are compatible with Lego. Once built, *Robo Wunderkind* can be easily programmed using a smartphone or tablet. Possible actions include driving around and avoiding obstacles, recording and playing voice messages, solving mazes, and even giving a weather forecast. The toy is also compatible with MIT's Scratch, a simple programming language recommended for ages eight and up, which allows children to expand the functionality of their robots even further. AR

Robo Technologies. *Robo Wunderkind,* 2015. Modular robotic toy, various materials and sizes © Robo Technologies GmbH

... and which comes with a Lego attachment.

Forbidden Planet is a 1956 American science fiction film noted for pioneering several aspects of science fiction cinema, including depicting humans travelling in a faster-than-light starship of their own creation, and the first to be set entirely on another planet. The story had wonderful characters, including "Robby the Robot"; for the first time ever in film, a robot displayed a distinct personality and was an integral supporting character. One of the great science fiction films of the 1950s, it also influenced many future Hollywood robots, droids, and other assorted mechanical men. It was nominated for the Academy Award for Best Visual Effects and in 2013 it was added to the National Film Registry of the Library of Congress. AR

FRED WILCOX
– *FORBIDDEN
PLANET*

MGM. *Forbidden Planet,* 1956.
Feature film, 98 min; director: Fred
Wilcox; film poster, 110 × 70 cm
© Warner Bros. Entertainment Inc.
All rights reserved

Forbidden Planet, film poster, 1956

The Matrix, 1999

Warner Bros. *The Matrix,* 1999.
Feature film, 126 min; directors
and script: the Wachowskis

THE WACHOWSKIS – *THE MATRIX*

In the 1999 sci-fi thriller *The Matrix,* the Wachowskis (Lana and Lilly Wachowski, screenplay and direction) tell the story of dystopian coexistence between humans and machines. In the earth's distant future, machines have become a highly developed species with artificial intelligence, thereby also becoming humanity's new masters.

People are bred to serve as nothing more than sources of energy, and the majority of them live in an induced state of sleep in giant breeding factories. They are unaware that for generations they have served as mere "batteries", for the machines deceive them by providing a life in the virtual world of illusion known as the Matrix. LH

53

After a five-year break from television appearances, the band Kraftwerk returned to the stage in 1978 with the single *Die Roboter* ("The Robots"). They performed as a robot band, wearing matching costumes and operating their electronic musical equipment with mechanical movements. Even before their comeback, the band was well known for its machine aesthetics, such as a "cold" style of singing devoid of all forms of emotion, as well as for the use of vocoders. Kraftwerk's transformation from the living to the synthetic was achieved by their introduction of lifelike plastic effigies (seen in the video), which the band would regularly use again in future performances. Ultimately, after all, the performance of electronic music is not contingent on the physical presence of the musician. TT

KRAFTWERK – *DIE ROBOTER*

Die Roboter, 1978

Kraftwerk. *1 2 3 4 5 6 7 8, THE MIX,* 2013. Video installation © Kraftwerk, courtesy Galerie Sprüth Magers

The original *The Ghost in the Shell* manga centres on a fictional anti-cyber-terrorism unit, Public Security Section 9. The protagonist is a female cyborg, Major Mokoto Kusanagi, who chases the Puppeteer, a cyber-criminal and hacker capable of taking over human minds.

The story is set in Japan in the year 2029, when nano-technology and robotics are everywhere, and cybernetic implants are widespread in society at large. The degree of body enhancement with electronics ranges from a simple interface between the brain and computer networks to various types of prosthetic replacements and even the complete replacement of the brain with a cyber-equivalent that enables humans to become cyborgs. The Puppeteer's main pursuit is "ghost hacking" or attacks on human brains. Major Kusanagi later discovers that the Puppeteer is an autonomous artificial intelligence project created by another government agency.

The Ghost in the Shell was made into a film several times and ranges among the most prominent Japanese science fiction *anime*. AR

MASAMUNE SHIROW
– *THE GHOST IN THE SHELL*

The Ghost in the Shell, manga, 1989

Masamune Shirow. *The Ghost in the Shell,* 1989. Manga © Masamune Shirow / Kodansha Ltd., courtesy Egmont Verlags-gesellschaften mbH

In the second strip on the page, Motoko Kusanagi, a cyborg and the protagonist of *The Ghost in the Shell*, discusses with a cyborg friend where to draw the line between humans, cyborgs and robots:
Friend: "Hey, we've got grey matter, and people treat us like humans…"
Motoko Kusanagi: "How do you know? You've never seen your alleged grey matter… Maybe you're just assuming you've got it because of the situation you're in. Maybe someday your 'Maker' will come haul you away, take you apart, and announce the recall of a defective product. What if all that's left of the 'Real You' is a couple of lonely brain cells, huh?"

MGM. *The Terminator,* 1984. Feature film, 107 min; director: James Cameron, script: James Cameron, Gale Anne Hurd © 1984 Metro-Goldwyn-Mayer Studios Inc. All rights reserved

The Terminator, 1984

JAMES CAMERON – *THE TERMINATOR*

The year is 2029. The Earth has been laid waste by a nuclear war fought between intelligent machines against their human creators. The humans, however, continue to fight a war of resistance led by John Connor, leading the machines to send the mechanical being known as the "Terminator" back in time to 1984 to kill – or to use the film's more euphemistic term, "terminate" – Connor's mother before she can give birth to the troublemaker. James Cameron's science fiction film envisions the killer robot as an efficient electronic machine covered in human tissue that allows it to move unnoticed in human society. The muscular *Terminator* with his super-hero stature was played by the Austrian bodybuilder and actor Arnold Schwarzenegger, whose slightly Styrian-accented "I'll be back" was to go down in cinematic history. TT

The Jetsons, 1962/63

An early 1960s vision of life in the twenty-first century was provided by the popular cartoon comedy series *The Jetsons,* which first appeared in 1962. In the series, every amenity is available at the touch of a button, people travel in flying saucers, and dinner is served by Rosie the robot housekeeper. Like their earlier series *The Flintstones,* this production by William Hanna and Joseph Barbera centres on an "average" American family and its day-to-day problems. Even though the *Jetsons'* fully automated version of the future is not without its pitfalls, the series reflects the unshaken faith in the progress of technology and its positive uses that is typical of the era. TT

57

HANNA-BARBERA – *THE JETSONS*

Universal Pictures. *Ex Machina,* 2015.
Feature film, 108 min; director and
script: Alex Garland © 2014 Universal,
Film4 and DNA Films

ALEX GARLAND – *EX MACHINA*

The British film *Ex Machina* tells the story of Caleb, a programmer who is invited by his boss to a secret research station where he is to test the female android Ava and determine whether her faculty of thought is equal to that of humans. Ava engages Caleb in intelligent discussions and manages to convince him of her own intelligence. After the two develop an emotional relationship, Ava ultimately persuades Caleb to help her escape from the research station. The film poignantly poses the question: if the simulation is perfect, does the distinction between humans and machines become irrelevant? TT

58

Ex Machina, 2015

Sleeper, 1973

MGM. *Sleeper,* 1973. Feature film,
88 min; director: Woody Allen,
script: Woody Allen, Marshall
Brickman © 1973 Metro-Goldwyn-
Mayer Studios Inc. All rights reserved

WOODY ALLEN – *SLEEPER*

In the science fiction parody *Sleeper,* the New York health-
food shop owner Miles Monroe – played by director
Woody Allen – is frozen in 1973. When he is thawed
out some 200 years later, he awakens to a dictatorship in
which people are completely controlled by the system:
they are subjected to brainwashing, are "programmed"
like computers, and have little more freedom than their
"labour-saving devices", their robot domestic servants.
Miles Monroe is the world's only free human being who
has yet to be brought to heel. To protect himself from
persecution, Monroe first pretends to be a robot, but in
the end he aids the rebels – the people who thawed him
out – in their victory. TT

I, Robot, 2004: good ...

The 2004 Will Smith blockbuster *I, Robot* reawakened the public's awareness of Isaac Asimov's eponymous short story collection, even if it did so indirectly. The screenplay, originally written by Jeff Vintar without any connection to Asimov's work, underwent gradual changes to incorporate references to Asimov's characters like Dr Susan Calvin (played by Bridget Moynahan), and later also to plotlines such as the police/robot mystery (personified by Will Smith and the robot, Sonny) from the novel *The Caves of Steel.* However, the most overt link to Asimov is the use of the "Three Laws of Robotics" as the film's ethical framework. Set in 2035, *I, Robot* depicts a world that has too hastily accepted new technology, even entrusting robots with every sphere of daily life, without considering its repercussions. EP

60

Twentieth Century Fox. *I, Robot,* 2004. Feature film, 115 min; director: Alex Proyas, screenplay: Jeff Vintar, Akiva Goldsmith; digital print © courtesy Twentieth Century Fox. All rights reserved

... or evil?

ALEX PROYAS – *I, ROBOT*

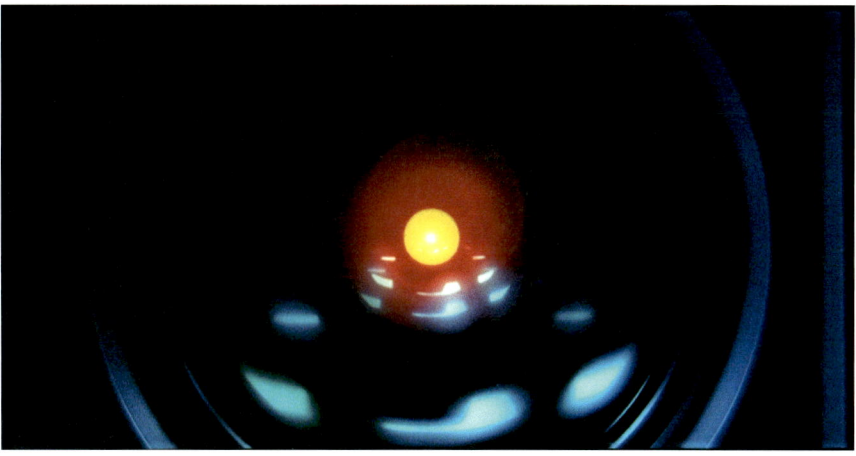

2001: A Space Odyssey, 1968

Stanley Kubrick's 1968 film is today considered one of the most famous science fiction films of all time. It is based on a short story by Arthur C. Clarke, who also helped develop the screenplay. During a mission in space, the supposedly infallible computer HAL 9000 reports a false diagnosis. When the two astronauts on board contemplate shutting it down, the computer rebels and attempts to stop them. In fear of being deactivated, HAL displays behaviour that increasingly resembles human emotions, to the point of inverting the classical stereotype.

HAL's interface, including the red eye and characteristic voice spoken in the original version by Douglas Rain, has become iconic in its own right. References to HAL 9000 can be found in countless films, books, and computer games. LH

STANLEY KUBRICK, ARTHUR C. CLARKE – *2001: A SPACE ODYSSEY*

GEORGE LUCAS – *R2-D2*

R2-D2 is a fictional robot from the American director George Lucas' *Star Wars* films. In the story, astromech droids such as *R2-D2* serve as mechanics with highly intelligent computing capabilities for repairing starships. Measuring 96 cm in height, "Artoo-Detoo" can communicate only by means of beeps, which are translated for the benefit of the audience by his humanoid robot partner and friend *C-3PO,* who understands millions of languages. *R2-D2* was played by the short-statured British actor Kenny Baker inside the robot's casing. However, in some scenes *R2-D2* was moved by remote control or computer animation. *R2-D2* and *C-3PO,* who have both appeared in all of the *Star Wars* films to date, are representative of typical fictional robots with superhuman capabilities. TT

George Lucas. *R2-D2,* 1977. First appearance in the feature film *Stars Wars: Episode IV – A New Hope* (1977); photograph © & ™ 2017 Lucasfilm Ltd. All rights reserved. Used under authorization.

R2-D2 and *C-3PO,* 1977

Hasbro and Takara. *Transformer,*
1984. Toy, various materials,
ca. 25 × 25 × 10 cm © Hasbro

Transformer, RID 3-Step Changer

The robot action figures have been produced by Hasbro and Takara since 1984 and can be transformed with just a few flicks of the wrist into cars, motorcycles, airplanes or everyday objects such as wristwatches, telephones, and radios. Their very name is derived from this ability to transform. While looking for new ways of marketing the toys, the manufacturers developed an animated series, *Transformers,* which introduced individual characters and their special characteristics. In addition to the television series, several films for the big screen, computer and video games, and countless merchandising articles have been produced. LH

Transformer, Steeljaw

HASBRO AND TAKARA
– *TRANSFORMER*

The French automobile manufacturer Citroën took advantage of the public's fondness for robots and transformers when they launched the Citroën C4 compact car in 2004. On this occasion, the company hired the London advertising agency Euro RSCG (now Havas Worldwide) to create a dancing C4 Transformer, which becomes "Alive with Technology" just as its campaign slogan suggests. Animated with the CG skills of director Neill Blomkamp and the moves of Justin Timberlake's choreographer Marty Kudelka, the car awakes at the sound of electronic music and dances around an empty parking deck. By showing a robot – consistently a powerful symbol in advertising – in the ad, Citroën demonstrated to customers the sheer quantity of new technology available in the C4 model. EP

The Citroën C4 transforms itself ...

... into a dancing robot ...

CITROËN – *DANCING CITROËN CARBOT*

Citroën and Euro RSCG London. *Dancing Citroën Carbot,* 2004. TV commerical, 29 sec © Citroën

... who knows how to shake its hips.

Knight Rider, 1982–1986

GLEN A. LARSON – *KNIGHT RIDER*

A man and his car in the fight against injustice. For four years from 1982 to 1986, Michael Knight (aka David Hasselhoff) and his intelligent car K.I.T.T. kept audiences glued to their screens. Thanks to turbo boost and super pursuit mode, K.I.T.T. (short for "Knight Industries Two Thousand") was not only faster than conventional automobiles, but it could also think, speak, drive by itself, jump over obstacles, drive on water, break open locks, and much, much more. Its molecular bonded shell made it indestructible, and to top it all off, it was the epitome of wit and charm. The American television series *Knight Rider* not only launched Hasselhoff's international career, but soon achieved cult status. It is still on television in several countries over thirty years after it was initially aired. LH

65

The song *Harder, Better, Faster, Stronger* was released on the Daft Punk studio album *Discovery* in 2001. The song's music video is a segment in the animated science fiction film, *Interstella 5555: The Story of the Secret Star System,* which was conceived as the visual realisation of the album by Daft Punk. *Interstella* follows the kidnapping of an alien rock band by an evil gold-digger who brings them to earth to exploit their talent. In the video segment for "Harder, Better, Faster, Stronger", the aliens are turned into androids and their memories are saved to disks. The film's robot imagery fits together perfectly with Daft Punk's trademark "robot" stage personas. EP

DAFT PUNK – *HARDER, BETTER, FASTER, STRONGER*

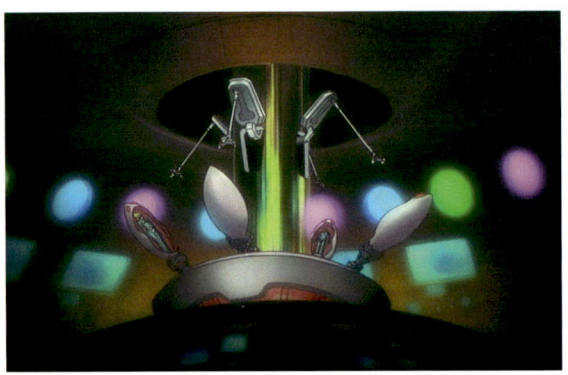

Harder, Better, Faster, Stronger, 2003

Daft Punk. *Harder, Better, Stronger,* 2003. Music video, 3 min 42 sec; director: Leiji Matsumoto © Daft Life Ltd under exclusive license to Parlophone / Warner Music France, a Warner Music Group company

ABB Ltd. *Spot Welding at Volvo
(with Integrated Dress Packs)*, 2014.
Video, 4 min 25 sec © ABB Ltd.

Spot Welding at Volvo, 2014

The first industrial robot was produced by Uni-mation, an American company founded by the engineer and entrepreneur Joseph Engelberger in 1961; it was first put into service at a General Motors factory. In 1974, ABB pioneered the first microprocessor-controlled industrial robot. Today, the company is one of the largest pro-ducers worldwide in the field of industrial robot-ics. Since the 1960s, the number of industrial robots worldwide has risen rapidly, with ap-proximately 260,000 units sold in 2015 alone. In the same year, China was recorded as the biggest importer of industrial robots, followed by Japan, the United States, and Germany. In 2013, ABB secured an order of 1,200 indus-trial robots to be delivered to Volvo factories in Torslanda and Olofström, Sweden: one of the factory assembly lines is depicted in the video. EP

ABB ROBOTICS – *SPOT WELDING AT VOLVO*

YVES GELLIE – *HUMAN VERSION*

For his photo series *Human Version,* French photographer Yves Gellie explored the biggest and most important research labs around the world specialising in humanoid robots. Gellie isn't just interested in the equipment and tools the labs use, but also in the spatial environments in which scientists apply themselves to their anthropoid machines. The focus on humanoids allowed Gellie to examine the interplay between scholarship (science) and pop culture (fiction) and explore the fundamental difference between East and West in attitudes towards robots. TT

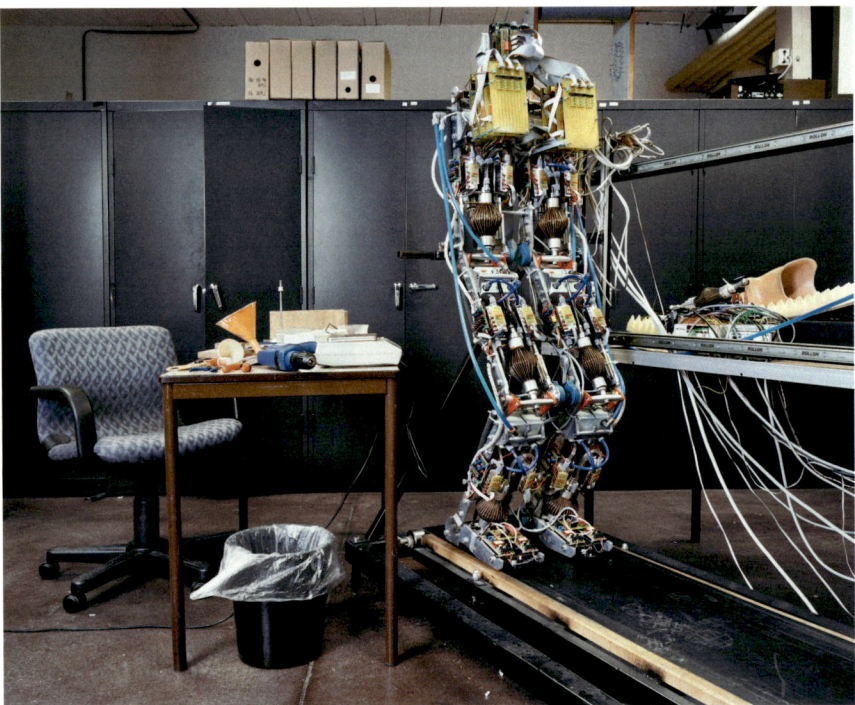

Human Version 2.03, Belgium, 2008: bipedal walking robot Lucy, developed by the Robotics and Multibody Mechanics Research Group at Vrije Universiteit Brussel (VUB).

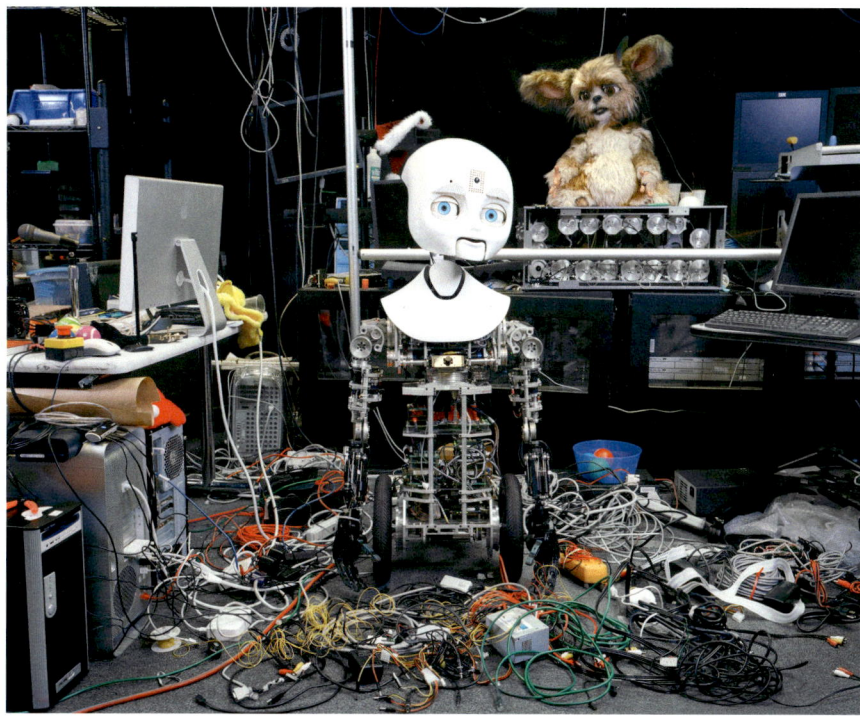

Human Version 2.07, USA, 2009: social robot Nexi, developed by the Personal Robots Group at the Massachusetts Institute of Technology (MIT).

Yves Gellie. *Human Version 2.03,* 2008. *Human Version 2.07,* 2009. *Human Version 2.010,* 2009. C-prints on baryta paper, 104 × 130 cm each © Yves Gellie galerie du jour agnès b, galerie baudoin lebon

Human Version 2.010, Japan, 2009: android robot Geminoid (l.), developed by Hiroshi Ishiguro (r.) in the Intelligent Robotics Laboratory at the University of Osaka.

ARE ROBOTS OUR FRIENDS OR
OUR ENEMIES?

WHAT WAS YOUR FIRST EXP
WITH A ROBOT?

DO WE REALLY N

HAVE
A ROI

DO YOU TRUST ROBOTS?

NCE

ED ROBOTS?

OU EVER MET

T?

Inspired by real events, the artist often asks sitters to im-mediately recreate their gestures, but this time without their personal device. The subjects stare at their hands or the empty space between their hands where a phone or tablet might have been. This pose makes it blatantly ob-vious that the people are often ignoring their immediate surroundings or opportunities for human connection.

Two scenes from the series are a couple in bed right before turning out the lights, (the photographer and his own wife), and three young boys on a summer afternoon. *Removed* aims to make our addiction to technology and hyper-connectivity tangible, something no one could have imagined just ten years ago. AR

ERIC PICKERSGILL – *REMOVED*

Eric Pickersgill. *Angie and Me; Grant; Head On* from the photo series *Removed,* 2014. Archival pigment print, 81.28 × 101.6 cm © 2014 Eric Pickersgill, courtesy Rick Wester Fine Art, New York

Angie and Me, 2014

Grant, 2014

Head On, 2014

NEXT NATURE NETWORK – *PYRAMID OF TECHNOLOGY*

The *Pyramid of Technology* is a model for describing the various levels at which technology functions in our lives. Movement from the base to the top of the pyramid describes the "naturalisation" of technology whereby new technologies transition from being perceived as artificial to feeling like a natural part of our lives. The stages are: Envisioned (such as the quantum computer), Operational (a prototype exists but is not widely applied, e.g. commercial space travel), Applied (it has moved out of the lab and into society, like chemotherapy), Accepted (an integrated part of daily life, like the light bulb), Vital (its disappearance would cause a lifestyle-changing crisis for its users, such as the Internet or sewage systems), Invisible (no longer recognised as technology, such as agriculture) and finally Naturalised (technology that we start to experience as second nature, like controlled fire). OP

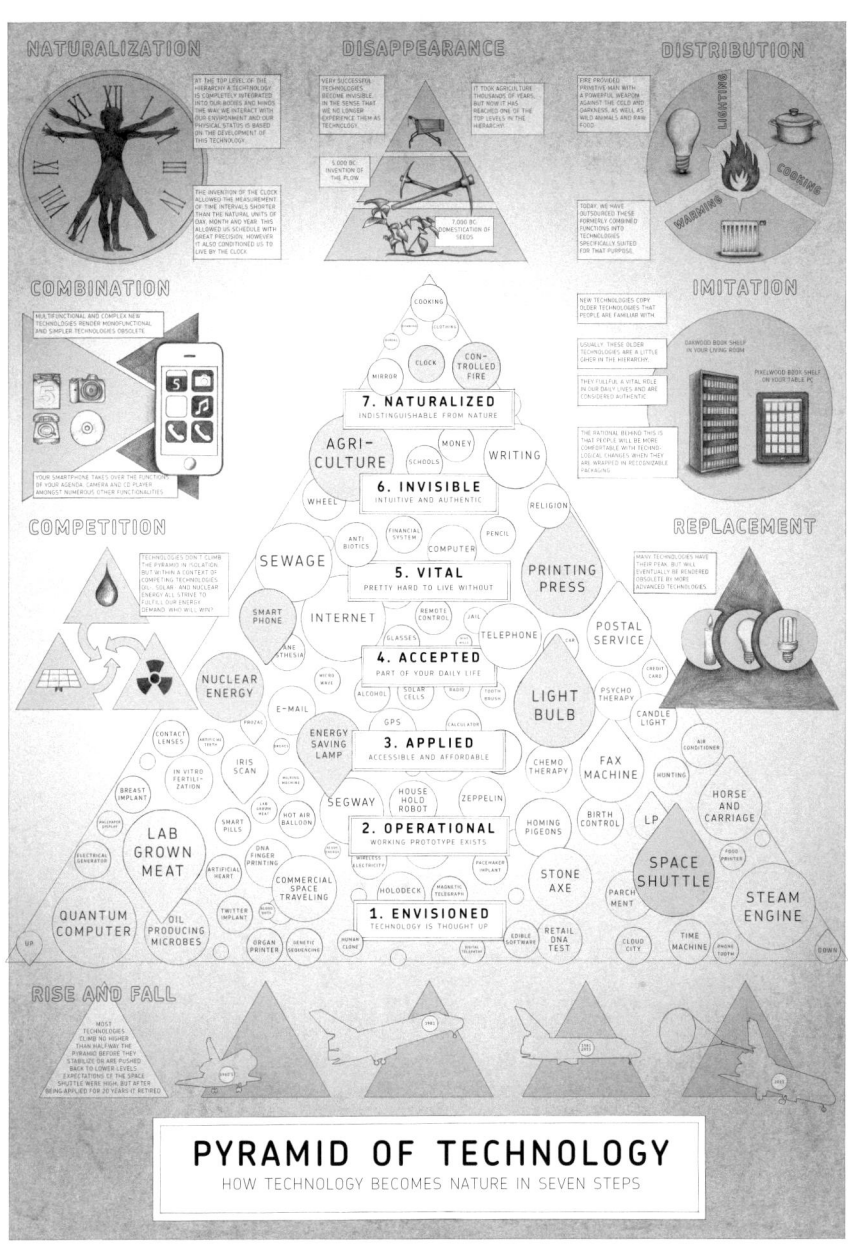

Pyramid of Technology, 2012

75

ARE ROBOT
OR OUR EN

HAVE YOU EVER MET A ROBOT?

DO YO

OUR FRIENDS
MIES?

WHAT WAS YOUR FIRST EXPERIENCE WITH A ROBOT?

DO WE REALLY NEED ROBOTS?

UST ROBOTS?

KEIJI INAFUNE, CAPCOM – *MEGA MAN*

Mega Man 5, 1992

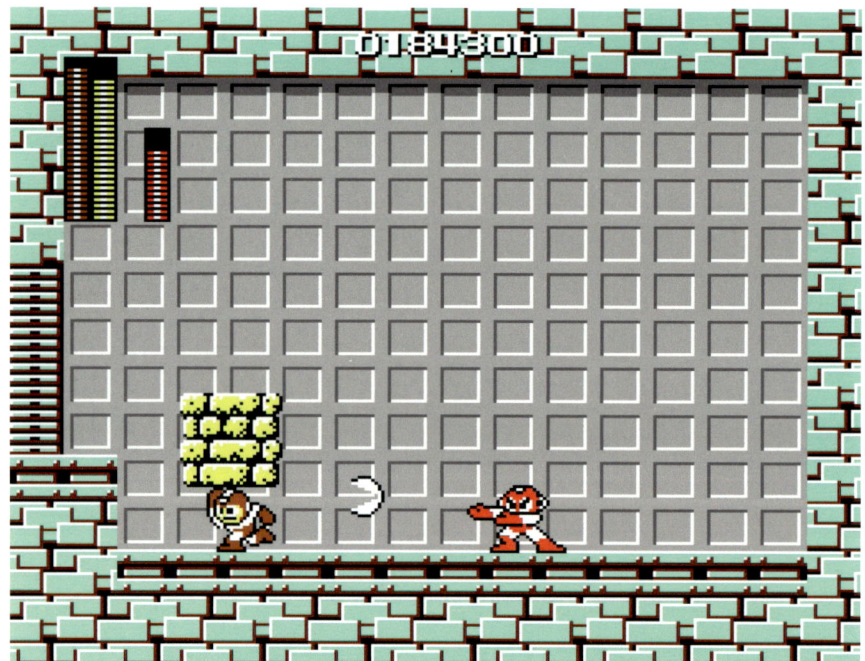

Keiji Inafune for Capcom. *Mega Man*, 1987. Video game for Nintendo © 2016 Capcom

Mega Man 1, 1987: The never-ending struggle between good and evil

The Japanese-made video game *Mega Man* was originally created by Capcom for Nintendo in 1987 and released as *Rockman*. Known for its difficulty and innovative graphics, it is considered a classic by the gaming community. The game's Western release plot (which deviated slightly from the Japanese original) is based on the seven humanoid creations of the scientists Dr Light and Dr Wily. Initially created to be beneficial to society, six of the robots are later re-programmed by Dr Wily to assist in his evil quest for world domination. Dr Light enlists the help of the last remaining robot, Mega Man, who moves through the various realms of "Monsteropolis", destroying countless enemies and obstacles before fighting each of the reigning evil robots. The game reflects society's ongoing questions about whether robots are working for us or against us – or both. EP

On 2 September 2015, a young Syrian boy named Aylan Kurdi drowned while his family was trying to cross the Mediterranean Sea to Europe. The photo of his small, lifeless body sparked an international outcry about migrant smuggling brought on by the Syrian refugee crisis. Chilean artist Alfredo Jaar used an image of the very beach where Kurdi died for his recent work *The Gift,* which calls on people to donate to "MOAS", the Migrant Offshore Aid Station. The Malta-based non-profit organises drones that scout the Mediterranean for migrant vessels in distress, ultimately sending search and rescue teams if necessary. Although drones are often associated with warfare and the negative connotations of surveillance, the MOAS initiative demonstrates how they can also help with countless search and rescue efforts, necessitated by either natural or political crisis. Jaar's emotional paper cube *The Gift* opens up to reveal the printed words "this gift can change you"; it might just change your view of drones too. EP

ALFREDO JAAR
– *THE GIFT*

79

The Gift, 2016

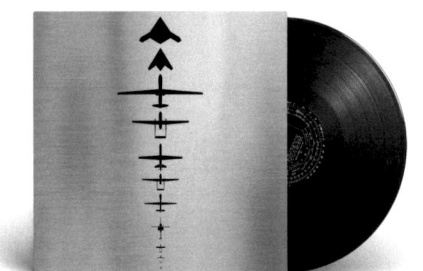

A Study into 21st Century Drone Acoustics, 2015: the sounds of 17 different drones

GONÇALO F. CARDOSO AND RUBEN PATER – *A STUDY INTO 21ST CENTURY DRONE ACOUSTICS*

Gonçalo F. Cardoso collaborated with designer Ruben Pater to create *A Study into 21st Century Drone Acoustics,* an audio guide to modern drones. The vinyl LP is supplemented by a 12-page insert including lists of drones, illustrations, maps, and essays. Although the imagery of drones has been stamped indelibly into the general public's awareness, most people are not familiar with the sounds they make or with the psychological effects of these sounds in conflict areas. For side A of the LP, the sounds of a variety of existing drones were recorded, such as the ScanEagle, the Predator MQ-I, and the Heron. For side B, Cardoso created a rousing original soundtrack, which was inspired by the abusive and destructive power of drone technology. EP

A Study into 21st Century Drone Acoustics, 2015: 12-page booklet

Gonçalo F. Cardoso and Ruben Pater. *A Study into 21st Century Drone Acoustics,* 2015. Audio installation with 12" LP record © Gonçalo F. Cardoso, Ruben Pater

Dragon Runner is a military robot developed in 2003 by Carnegie Mellon University in Pittsburgh in partnership with the manufacturer Automatika for the U.S. Marine Corps. It was used for the first time during the Iraq War, aka "Operation Iraqi Freedom". The robot is very light, rugged, compact, and portable. It will fit in a backpack and can be controlled by soldiers from a safe distance; it can navigate multiple terrains; it can help experts find and deactivate bombs without endangering soldiers; it can send video footage and audio back to the operator, thus providing information for troops before entering a potentially hazardous place. The robot is also modular and configurable so that enhancements, such as a manipulator's arm, can be added in the field as needed. In 2008, *Dragon Runner* made its most important pop culture appearance in the Oscar-winning film *The Hurt Locker*. AR

Dragon Runner in the photograph *Captain Judith Gallagher of 11 EOD (Explosive Ordnance Division)*, 2010

Geoff Caddick. *Captain Judith Gallagher of 11 EOD (Explosive Ordnance Division)*, 2010. Photo displaying an anti-IED robot known as the "Dragon Runner"; digital image © Geoff Caddick / AFP / Getty Images

CARNEGIE MELLON UNIVERSITY, AUTOMATIKA AND U.S. MARINE CORPS – *DRAGON RUNNER*

NEXT NATURE NETWORK – *WHAT'S FLYING THERE?*

What's Flying There? is a colouring book for children and adults published by Next Nature Network that creates another perspective on the controversial subject of drones. The book tells the story of a little bird named Avi that lives in a war zone and is hunted down by military drones. One day, Avi escapes to a place where drones do good things, like farming the land and fighting diseases. The book imagines the potential that exists between the extremes of drones as killing machines and drones as an innovative new parcel delivery system. OP

What's Flying There?, 2016

Avi the bird tries to escape from a military drone, ...

... meets some Wi-Fi drones, ...

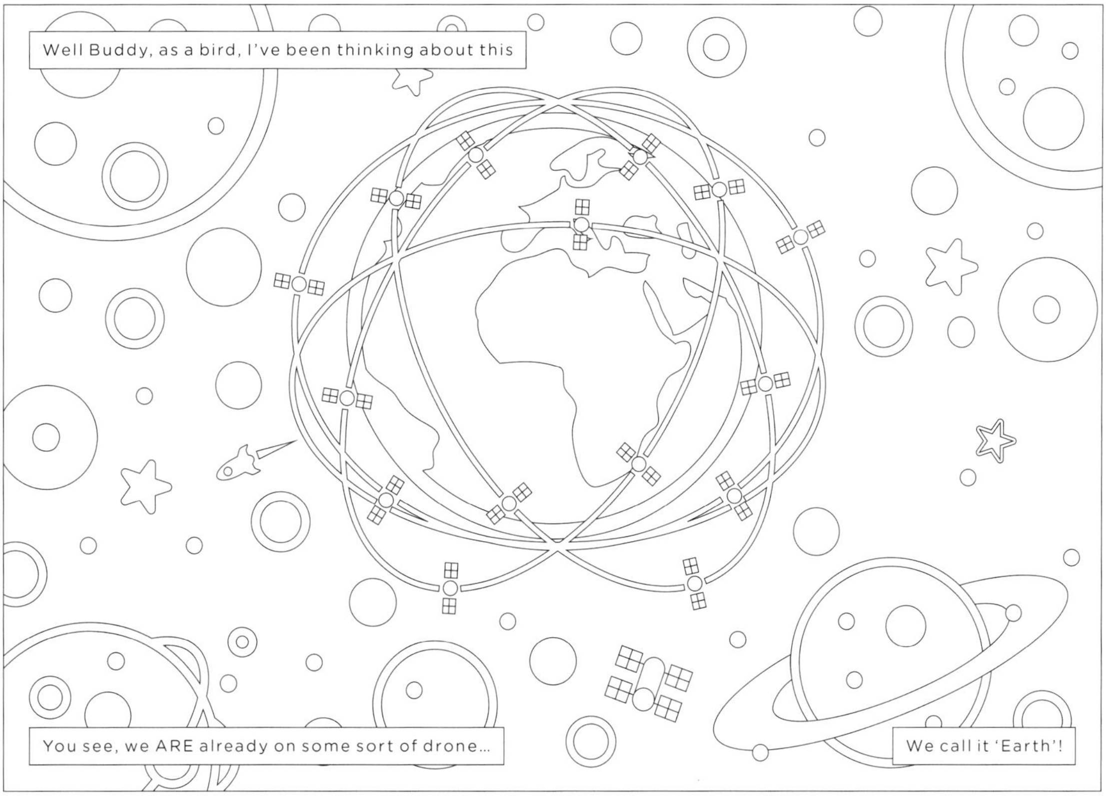

... and realises that the whole world is basically a drone.

Next Nature Network. *What's Flying There?,* 2016. Colouring book, 32.5 × 23 cm; creative direction: Koert van Mensvoort, art direction: Hendrik-Jan Grievink; story: Hessel Hoogerhuis, Hendrik-Jan Grievink; sound design: Arnoud Traa; communication: Lotte Mertens © 2016 Next Nature Network

In his photo series *The Man Machine,* Vincent Fournier captures speculative scenes featuring everyday encounters between humans and robots, their interactions, and empathic moments. The photos look at the degree to which robots are accepted by humans; while one might assume that the more realistic their appearance, the more robots find acceptance, in reality, the opposite is the case. According to the *Uncanny Valley* theory of Japanese roboticist Masahiro Mori, people find abstract, artificial-looking human replicas more attractive than those that resemble us too closely. When a certain degree of similarity is reached, acceptance decreases abruptly and only increases when the level of similarity with real human beings is extremely high or the differences are indistinguishable. TT

VINCENT FOURNIER
– *THE MAN MACHINE*

Reem B #6 [Pal], Barcelona, Spain, 2010

Vincent Fournier. *The Man Machine*, 2010. *Reem B #6 [Pal], Barcelona, Spain; Reem B #7 [Pal], Barcelona, Spain; Reem B #5 [Pal], Barcelona, Spain.* Series of 21 photographs, ink jet prints, 100 × 130 cm each
© Vincent Fournier

Reem B #5 [Pal], Barcelona, Spain, 2010

Reem B #7 [Pal], Barcelona, Spain, 2010

ECAL – *THYMIO MEETS ECAL*

Thymio is a small robot equipped with wheels, multiple sensors, and actuators created in collaboration between ECAL (École cantonale d'art de Lausanne) and EPFL (École Polytechnique Fédérale de Lausanne). It was designed primarily as a teaching aid for introducing children and young people to robotics and robot programming. In the video shown, students of Media & Interaction Design at ECAL put the robot's skills to the test and made it "write" words in various ways. The experience taught them that the good (and simultaneously the bad) thing about robots is that they do what you tell them to. This also means that robots are only as dangerous as the decisions made by human individuals and societies. TT

ECAL University of Art and Design, Lausanne, BA Media & Interaction Design. *Thymio Meets ECAL*, 2015. Video with Thymio robots (EPFL / ECAL / ETHZ / Mobsya), 2 min 5 sec; tutor: Alain Bellet; assistants: Laura Perrenoud, Tibor Udvari (ECAL), Manon Briod, Maria Beltran (EPFL); video: Arthur Touchais © 2016 ECAL

Thymio Meets ECAL, 2015

Kill Your Co-Workers, 2010

Flying Lotus. *Kill Your Co-Workers,*
2010. Music video, 3 min 4 sec,
direction and animation: Beeple aka
Mike Winkelmann © Warp Records,
2017, photos: Mike Winkelmann

A cheerful parade ...

... turns into a massacre ...

This music video is the result of a successful collaboration
between the musician Flying Lotus and the video artist
Beeple. Fitting with the EP's title, *Pattern+Grid World,*
the music video is set in a world of animated, geometric
forms. We see a wondrous parade of people and robots
waving happily and marching along to the beat of Flying
Lotus' synthesiser. The cheerful mood continues even
after the robots suddenly begin to brutally slaughter the
humans. Heads and body parts roll, cube-like drops of
blood spray forth, and the maimed and murdered victims
just keep cheering. The gem of a music video shows a
cynical and absurd coexistence between humans and machines. LH

87

... then things go right back to normal.

FLYING LOTUS – *KILL YOUR CO-WORKERS*

SUPERFLUX – *UNINVITED GUESTS*

Uninvited Guests, 2015

Smart objects

Superflux. *Uninvited Guests,* 2015. Installation, various materials and sizes; video: 4 min 43 sec; commissioned for: ThingTank; team: Anab Jain, Jon Ardern, Jon Flint, Alexandra Fruhstorfer, Katarina Medic, James Leahy; acknowledgements: James Leahy, Prof. Chris Speed and the ThingTank consortium © Superflux

Uninvited Guests is a design fiction project that explores the flip side of the pervasive presence of digital gadgets in our lives and our relationship with them. The short film focuses on the idea of a "connected home", where imagined smart objects help to improve the quality of life of an elderly parent by monitoring his day-to-day activities. Tension arises when the tracking devices begin to feel less like caregivers and more like an annoyance by constantly policing the man's every move, from his eating habits to his exercise and sleep patterns. The film not only questions the power dynamic between technology and its human users, but also comments on how technology changes communications and relationships between people. AR

Manu Cornet. *Mobile Relationship*, 2012. Graphic print © Manu Cornet

Mobile Relationship, 2012

The cartoons of the French illustrator and programmer Manu Cornet allude to the ambivalent relationship between modern humans and their constant companions: digitally networked smartphones. Whereas we do indeed believe that these devices are labour-saving and useful, the reality is – according to Cornet – that we are their slaves. They incessantly demand we react and communicate, and many of the tasks they apparently perform on our behalf in our networked lives were introduced by the smartphone in the first place. The question whether technology is a friend or an enemy ultimately comes down to the question of who is controlling whom. TT

89

MANU CORNET
– *MOBILE RELATIONSHIP*

DO YOU TRUST ROBOTS?

WHAT WAS YOUR FIRST EXPERIENCE WITH A ROBOT?

?

HAVE YOU EVER MET
A ROBOT?

ARE ROBOTS
OUR FRIENDS
OR OUR ENE-
MIES?

Women's Tech, an engineering association based in the Democratic Republic of the Congo, took the idea of *Robocop* to a new level with their introduction of traffic robots in 2013. The country's capital city Kinshasa purchased a series of the imposing robots, which are 2.5 m high and weigh 250 kg each, in order to improve the inefficient traffic system; while their human counterparts have been known to take bribes, the robots send all video (filmed via their eyes and shoulders) directly to the main station for ticketing and control, thereby gaining the respect of locals. Despite their high price tag of $27,500, the robots are solar-powered, and thanks to their aluminium plating, they can withstand long shifts in the hot Congolese sun. EP

WOMEN'S TECH – *TRAFFIC ROBOT*

Traffic Robot, 2015. Photo © Federico Scoppa / AFP / Getty Images

Al Jazeera America. *DR Congo
Recruits Robots as Traffic Police,*
2014. Video, 2 min 23 sec © 2016
Al Jazeera Media Network

knife.hand.chop.bot, 2007

5Voltcore. *knife.hand.chop.bot,* 2007. Mixed media installation, ca. 53.5 × 50 × 130 cm © Emanuel Andel

5VOLTCORE – *KNIFE. HAND.CHOP. BOT*

Whether you call it the "pinfinger", "stabscotch", or just the "knife game", rapidly stabbing the narrow places between your fingers is anything but comforting. Still, the stabber remains in control of his or her own destiny, or fingers, at least. Taking these tensions to the next level, the (since disbanded) artist collective 5Voltcore presented *knife.hand.chop.bot* in 2007 not merely as an outdated test of courage, but as an instrument to test our trust in technology. Each participant stretches out their hand and allows the robot to stab in between their fingers at a gradually faster pace. If the person gets nervous, their perspiration can potentially cause the otherwise programmed-to-precision "bot" to slip and make a big, uncomfortable mistake. But then again, maybe it wasn't a mistake after all. EP

94

JOSEPH POPPER – *WHEN THE HOME STOPS*

When the Home Stops is a speculative design project created by Joseph Popper. The video shows a lone man in the very familiar setting of his home, where he has entrusted even the most intimate tasks, such as flossing his teeth, shaving, and showering, to a domestic robot. The video shows an automated system that appears to have stopped functioning entirely, leaving the man, who has become completely dependent on his machines, helpless and panic-stricken. The series of "still lifes" explores what can happen if our trust in and dependence on artificial intelligence goes too far. AR

When the Home Stops, 2011

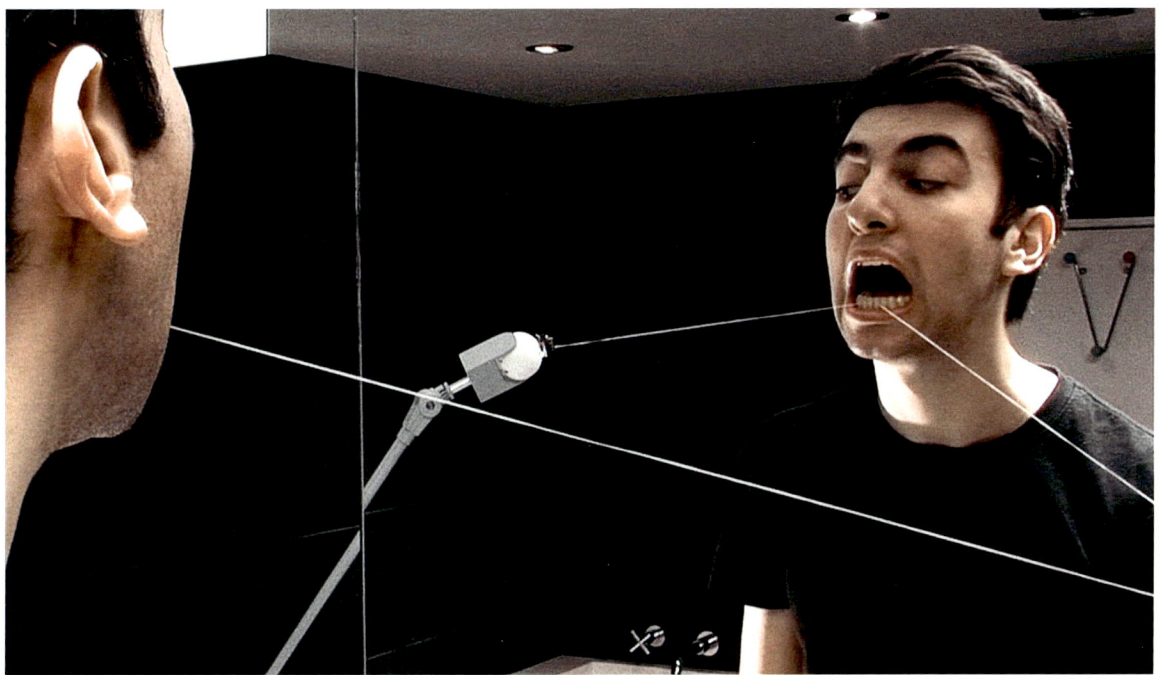

His dependence on robots ...

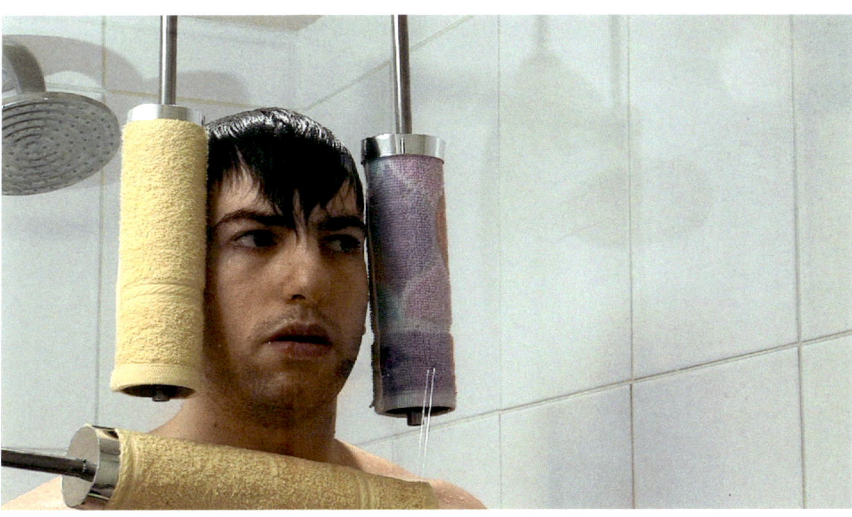

... leaves the protagonist frozen in a state of complete helplessness.

Digicars are part of a design fiction by British designers Dunne & Raby. In the imaginary countries of the "United Micro Kingdoms", these self-driving vehicles are the means of transportation of the Digitarians, a society governed entirely by market forces. Therefore, Digicars don't just navigate from A to B; they also connect passengers as efficiently as possible with the best offers on the market. Unlike the idea behind the self-driving cars being developed today, Digicars are less a means of leisurely travel and more a relative of the budget airline – a highly marketised, fast-track mode of transportation where comfort often comes last. TT

NUMBER OF PASSENGERS	CAR SIZES	CABIN STYLES	PRIVACY	DROP-OFF PRIORITY	PRIORITY RIGHT OF WAY	STOPS PER DAY	PEAK USAGE	OFF-PEAK USAGE
1	S	1 FRONT FACING	MAXIMUM	1ST	5 STAR	1	8 HOURS	16 HOURS
2	M	1 REAR FACING	ENHANCED	2ND	4 STAR	2	7 HOURS	14 HOURS
3	L	1 NO WINDOWS	STANDARD	3RD	3 STAR	3	6 HOURS	12 HOURS
4		1 RECLINER		4TH	2 STAR	4	5 HOURS	10 HOURS
		2 FRONT FACING			1 STAR	5	4 HOURS	8 HOURS
		2 REAR FACING				6	3 HOURS	6 HOURS
		2 SIDE FACING				7	2 HOURS	4 HOURS
		2 NO WINDOWS				8	1 HOUR	2 HOURS
		3 FRONT FACING				9		1 HOUR
		3 REAR FACING				10		
		3 SIDE FACING				11		
		3 NO WINDOWS				12		
		4 FRONT FACING				13		
		4 REAR FACING				14		
		4 SIDE FACING				15		
		4 NO WINDOWS				16		
						17		
						18		
						19		
						20		
						21		
						22		
						23		
						24		

Digicars, 2012/2013: tariff chart

DUNNE & RABY – DIGICARS

Dunne & Raby. Digicars from the project United Micro Kingdoms, 2012/13. "Digicar", cardboard, ca. 54 × 35 × 35 cm; CGI rendering: Tommaso Lanza; tariff calculation: Tobias Revell; tariff illustration: Kellenberger-White © Dunne & Raby

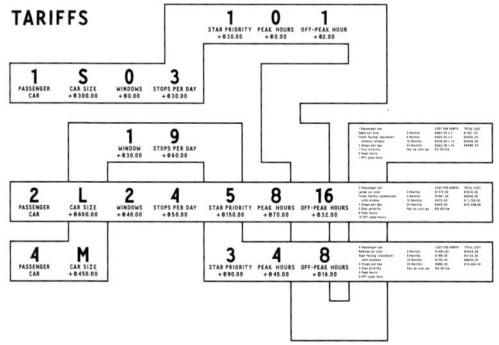

TARIFFS

Whoever pays more gets ahead first.

Self-driving cars ...

... as a highly marketised mode of transportation ...

... comparable to budget airlines. Photo © Luke Hayes / Design Museum

97

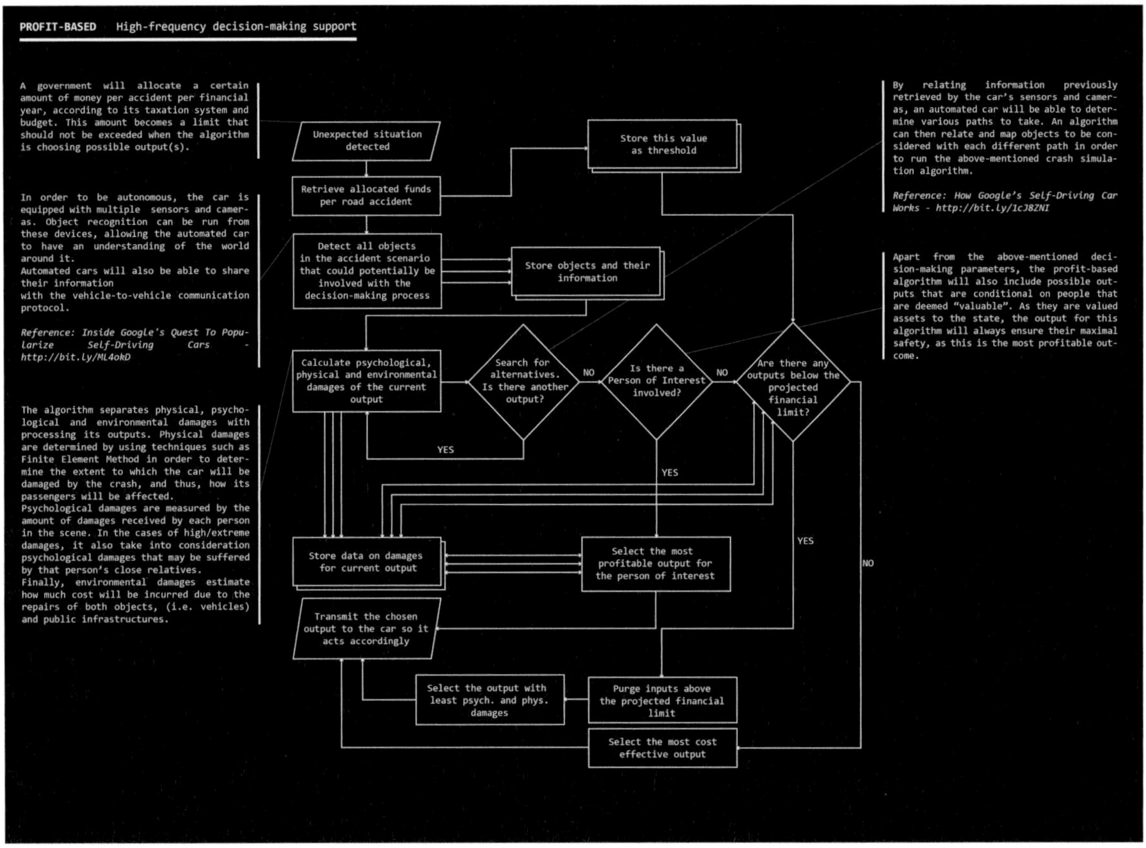

PROFIT-BASED High-frequency decision-making support

A government will allocate a certain amount of money per accident per financial year, according to its taxation system and budget. This amount becomes a limit that should not be exceeded when the algorithm is choosing possible output(s).

In order to be autonomous, the car is equipped with multiple sensors and cameras. Object recognition can be run from these devices, allowing the automated car to have an understanding of the world around it.
Automated cars will also be able to share their information with the vehicle-to-vehicle communication protocol.

Reference: Inside Google's Quest To Popularize Self-Driving Cars - http://bit.ly/ML4okD

The algorithm separates physical, psychological and environmental damages with processing its outputs. Physical damages are determined by using techniques such as Finite Element Method in order to determine the extent to which the car will be damaged by the crash, and thus, how its passengers will be affected.
Psychological damages are measured by the amount of damages received by each person in the scene. In the cases of high/extreme damages, it also take into consideration psychological damages that may be suffered by that person's close relatives.
Finally, environmental damages estimate how much cost will be incurred due to the repairs of both objects, (i.e. vehicles) and public infrastructures.

By relating information previously retrieved by the car's sensors and cameras, an automated car will be able to determine various paths to take. An algorithm can then relate and map objects to be considered with each different path in order to run the above-mentioned crash simulation algorithm.

Reference: How Google's Self-Driving Car Works - http://bit.ly/1cJ8ZNI

Apart from the above-mentioned decision-making parameters, the profit-based algorithm will also include possible outputs that are conditional on people that are deemed "valuable". As they are valued assets to the state, the output for this algorithm will always ensure their maximal safety, as this is the most profitable outcome.

Flowchart: Unexpected situation detected → Retrieve allocated funds per road accident → Store this value as threshold → Detect all objects in the accident scenario that could potentially be involved with the decision-making process → Store objects and their information → Calculate psychological, physical and environmental damages of the current output → Search for alternatives. Is there another output? — NO → Is there a Person of Interest involved? — NO → Are there any outputs below the projected financial limit? Store data on damages for current output · Select the most profitable output for the person of interest · Transmit the chosen output to the car so it acts accordingly · Select the output with least psych. and phys. damages · Purge inputs above the projected financial limit · Select the most cost effective output. (YES / NO branches)

Ethical Autonomous Vehicles, 2015

MATTHIEU CHERUBINI (AUTO-MATO.FARM) – *ETHICAL AU-TONOMOUS VEHICLES*

Matthieu Cherubini (automato.farm).
Ethical Autonomous Vehicles, 2015.
Video, 6 min 59 sec © automato.farm

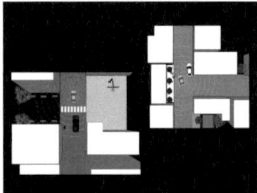

Accident scenario #2

Many people are predicting that self-driving vehicles will be safer than current forms of transportation. But even if this is true, these vehicles will also encounter moral and ethical dilemmas, such as when an accident is unavoidable and a decision has to be made as to who is injured and to what extent. The video *Ethical Autonomous Vehicles* simulates accidents on the basis of three different algorithms, each of which follow a particular ethical position: humanist (aimed at minimising social and human impacts), profit-based (aimed at minimising monetary cost), and protective (aimed at minimising danger to the passenger). Viewers can see how, in the end, such systems always reflect existing ideological positions. TT

AUTOMATO.FARM – *ETHICAL THINGS*

Ethical Things, 2015: the crowd decides.

Artificial intelligence still has difficulty making ethical decisions. The automato. farm collective asks how we can give everyday devices – such as fans, for example – the ability to make decisions that involve ethical dilemmas.

Which person in a room should be given preference when ventilating the room? Here the device draws upon a technique known as "turking", borrowed from Wolfgang von Kempelen's famous "Mechanical Turk".

Whereas von Kempelen's chess-playing automaton in fact contained a human chess player who operated the puppet dressed in Turkish traditional costume, the networked fan receives its answers from a crowdsourcing platform. The age, sex, education, and religion of the "turkers" is predetermined in order to come up with an ethical decision that meets the needs of the situation as well as the crowd.
TT

automato.farm (Simone Rebaudengo, Matthieu Cherubini and Saurabh Datta). *Ethical Things,* 2015. Fan, various materials, 25 × 8 × 10 cm; booklet, 159 pages © automato.farm

Ethical Things, 2015

The series of objects presented in the installation *Politics of Power* uses the simple example of three differently controlled multi-plugs in order to illustrate how products and networks are always influenced by ideologies and world views, including those of their designers and engineers. The networks of the three multi-plugs are organised differently: one is egalitarian, the second hierarchical, and the third is unequal and repressive – reflecting various social systems – and they behave in accordance with these models when distributing electricity. The demonstration helps us to think about the way we are dependent not only on our devices, but also on the resources that they require to function and on the people who control these resources. TT

automato.farm (Simone Rebaudengo, Matthieu Cherubini and Saurabh Datta). *Politics of Power*, 2016. Light installation; model M & D: 30 × 26 × 6 cm, model T: 32 × 32 × 6 cm
© automato.farm

Politics of Power, 2016: Distribution of power...

... according to different social models

AUTOMATO.FARM – *POLITICS OF POWER*

Superflux. *The Drone Aviary,* 2015.
Installation with three drones, various materials and sizes; video, 6 min 34 sec; project lead: Jon Ardern and Anab Jain; design and prototyping: Jon Flint, Jon Ardern, Dillon Froelich, Ian Hutchinson, DOME Studio; script and direction: Anab Jain; visual design: Katarina Medic, Georgina Bourke; motion design: Dimitris Papadimitriou, Laurence Mencé, Alexandra Fruhstorfer; sound design: Sam Conran, Ian Rawes, Gwaith Swn, London sound survey; technologists: Jon Ardern, Dan Williams, Mike Vanis, Philipp Ronnenberg; photos: Owen Richards, Jon Flint, Jon Ardern, Anab Jain; drone fictions: Tim Maughan; film footage: Traffiko, Geoweb3d, Phoenix Lidar System, Cedric Guillemet; acknowledgements: Yosuke Ushigome, David Benque, Elvira Grob, Gejin Gao, Anuradha Reddy, Tobias Revell, Carolina Vallejo, Marty Brown and Mariko Oya © Superflux

The Drone Aviary, 2015: Information?

SUPERFLUX – *THE DRONE AVIARY*

The Drone Aviary is a design fiction project which consists of an installation of five drones and a short film. For the project, Superflux has predicted a dystopian future where drones are as commonplace and easy to use as smartphones. They become protagonists in the public and private realm, as they constantly collect data and perform tasks in law enforcement, personal communication, news, and advertising. The scenario hints at a world where the "Net" begins to gain physical autonomy, moving within and making decisions about the world, influencing our lives in opaque yet profound ways. A speculative map highlights the points where physical and digital infrastructures merge while our cities become the natural habitat for "smart" technology such as drones, wearable computers, and driverless cars. AR

Protection?

Surveillance?

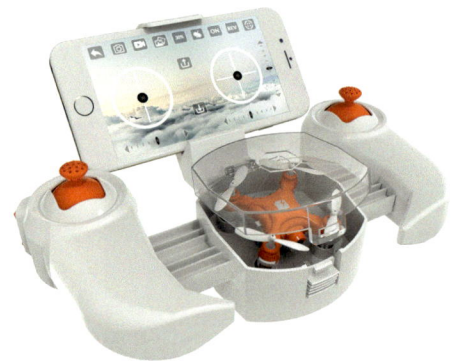

SKEYE Nano 2 FPV, 2016

Only three days after its initial release on 6 July, 2016, the augmented reality game Pokémon Go already had more users that the short message service Twitter, launched in 2006. By 12 July, 21 million Americans were hunting for virtual Pokémons, and on 13 July the Dutch company TRNDlabs announced the *Pokédrone,* a $69 drone for the smartphone that allows users to capture the Pokémons in inaccessible locations such as on the water – a marked advantage, for in order to move throughout the virtual world of the Pokémons, players have to use their phones to navigate through the real world. The drone was never launched on the market.

Pokémon Go also broke records in the negative sense. In the initial version of the game, users had to grant full access to all of the data on their phones. Even after this security flaw was repaired on 13 July, the game developer Niantic, a spin-off of Google, continued to collect sensitive data from millions of users* and made it available to third parties. For this reason, soldiers in the Israel Defence Forces are not allowed to use the app on military bases. The game allowed Nintendo, the Japanese game and console manufacturer that first brought the Pokémon videogames to market in 1996, to more than double its market value within seven trading days after the introduction of Pokémon Go.

* Daily users as of 18 July: 45 million; daily users as of 23 August: approx. 30 million (source: Bloomberg). Downloads via Google Play as of 8 August: 100 million; downloads via App Store: unknown (source: Business of Apps). LH

TRNDLABS – *SKEYE NANO 2 FPV (POKÉDRONE)*

ALEXANDER REBEN – *BLABDROID*

Alexander Reben. *Blabdroid,* 2012.
Art project, various materials, 14 ×
12 × 12 cm © Alexander Reben,
photo: Michael Underwood

"What's your name?" "Do you find me cute?"
With these simple questions, the pre-programmed
robot *Blabdroid* breaks the ice with complete
strangers who end up telling it secrets they have
never told anyone before and sharing embar-
rassing stories and other anecdotes. Equipped
with the husky voice of a seven-year-old boy and
a face dominated by a pair of wide-set eyes, the
minuscule *Blabdroid* exactly corresponds to our
idea of "cuteness", making it very easy for people
to open up to it. The little robots were created
as successors to "Boxie", Alexander Reben's thesis
project at MIT in the field of human-robot-
symbiosis. Reben then teamed up with filmmaker
Brent Hoff to create *Robots in Residence,* billed
as the first documentary to be "shot and directed
entirely by robots". Unsurprisingly, the results
are eminently human. EP

103

Blabdroid, 2012

The 8ᵗʰ Sphere, a wall chart devised by the research and design collective Bureau d'Études, represents an attempt to illustrate the communication channels and power relationships inherent in cognitive capitalism. In a world shaped by new information and communication technologies, all machines and computers have been joined together to create a giant network which covers the entire globe (in line with Marx's notion of a general intellect) and which, like a nervous system, is made up of synapses. As a socio-technical system, this network produces knowledge while at the same time defining the social and bio-political order. The human intellect too is embedded in this structure, and this in turn leads to the question of our minds' status. Are we undergoing a transition from organic to technical intellect, or has this transformation already taken place? TT

BUREAU D'ÉTUDES – *THE 8ᵀᴴ SPHERE*

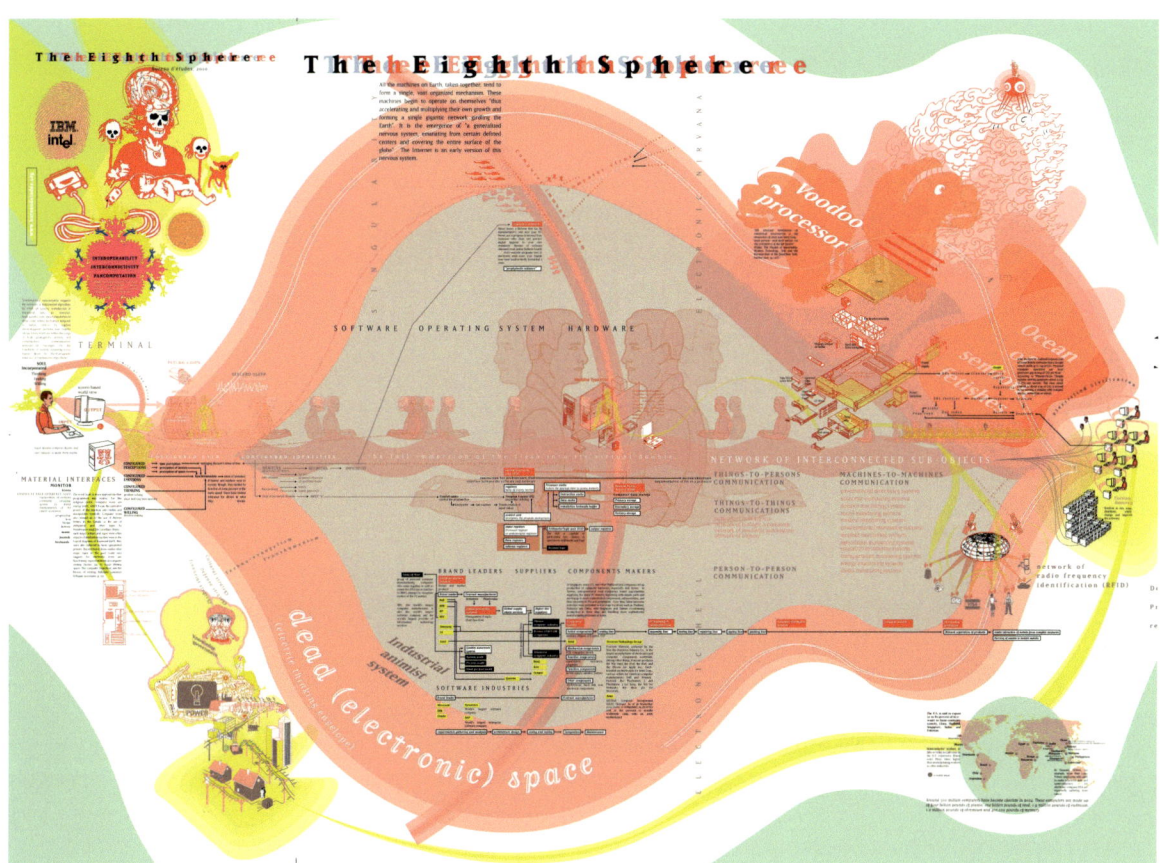

The 8th Sphere, 2010–2016

Bureau d'Études. *The 8th Sphere*,
2010–2016. Graphic art in light box
88 × 123 cm © Bureau d'Études

PROGRAMMED TO WORK

In the world of work, production, and industry – among the general public at least – robots are strongly associated with the fear of job loss. The issue is the subject of heated debate not only in the media; designers, artists, and filmmakers too are looking at what happens when people are gradually replaced by intelligent machines in the workplace. Will our standard of living decline along with our income? Or will we finally have more time for our friends, families, and hobbies thanks to a three-day working week and an unconditional basic income? Will new professions arise, and if so, what will they be? Will we work side by side with robots who are fully networked with customers and suppliers, as Industry 4.0 promises?

The fear of losing jobs to new technologies is as old as the first industrial revolution. In those days it was looms and steam engines that rendered hundreds of thousands of jobs obsolete. Since then, every technological leap has triggered the same discussions: most recently with the PC in the 1980s, the Internet in the 1990s, and now with robots. Time has shown that we have always found new areas of work even if working conditions have changed dramatically since the eighteenth century. *Hello, Robot.* seeks to shed light on the current debate from different perspectives against the background of technological and social change.

In the shadow of this debate, which ultimately still presupposes traditional structures of production and labour, a completely new breed of human has evolved: the prosumer. Prosumers consume what they produce themselves. What distinguishes them from individual world reformers is that they are globally networked via the Internet and have easy access to new, digital, robotic means of production. Both factors allow prosumers to sidestep traditional markets in order to develop, produce, and distribute tailor-made, smart products. Even today digital processes such as 3D printing make it possible for individuals to produce small pieces of furniture or everyday objects at a reasonable price. To produce a bridge, a house, or a haute couture dress obviously requires a wider range of competencies, but once designs and building plans are accessible to everyone online and open workshops and Fab Labs have become as commonplace as gyms, everyone will be able to produce (almost) everything themselves. Whether the end of the division of labour and the return to self-sufficiency will actually solve all our problems remains to be discussed.

ROBOTS, KILOBOTS, NANOBOTS

SWARMING IN THE INTERNET OF THINGS

By 2020, fifty billion objects will be connected via the Internet of Things,[1] far more than there are people on the planet to regulate or control them. Indeed, the Internet of Things is by no means just about computing in its original sense, but rather about such things as interlinking street lights with the urban traffic system and combining harvesters with sensors in the ground, or implanting chips in the body to analyse health data. The digitisation and miniaturisation of technology has given rise to new distributed systems that can no longer be clearly mapped but instead disseminate processing power everywhere as ubiquitous computing – in the networked house (smart homes), in intelligent clothing (wearable computing), and in public spaces (smart cities). The attempt to keep "nature" and "technology"

separate is becoming increasingly futile – for we are now dealing with hybrid, networked worlds. Indeed, Donna Haraway's manifesto,[2] for one, has made it abundantly clear that we have finally turned into cyborgs, a cross between human and machine. We are ourselves part of the Internet of Things, inasmuch as we use the technology surrounding us to analyse our personal data and share it online. Following Bruno Latour, we can argue that this kind of modern demarcation between nature and technology never existed; rather, it was always artificially produced to make a clean division and keep things simple.[3] After all, technology is increasingly becoming part of our "natural" body and the "natural" environment. As the debate about whether "robots" will take jobs away from us humans heats up, we should bear in mind that the problem is already inherent in the discrete categories we have set up: processing power

does not emanate simply from a clearly differentiated robot, which would be easy – rather it is about different actors in a network in which we become collaborators with the machines. Processing power can be found in us and around us. When our own job is under threat, "the robot" becomes a clearly definable entity that is perfectly suited to assume the mantle of "enemy", but a networked system involving new forms of collaboration between human and machine is not so easy to get to grips with. Thus, the system in Latour's actor-network theory does not care whether it is dealing with a human or a non-human actor.[4] In the financial services system, for instance, it is irrelevant for the process whether a decision has been made by an algorithm or a human being. An individual citizen whose loan application has been rejected on the basis of an algorithm is likely to see things differently.

1 Dave Evans, "The Internet of Things: How the Next Evolution of the Internet Is Changing Everything", http://www.cisco.com/c/dam/en_us/about/ac79/docs/innov/IoT_IBSG_0411FINAL.pdf, accessed on 24 October 2016.
2 Donna Haraway, "A Cyborg Manifesto: Science, Technology and Socialist-Feminism in the Late Twentieth Century", in David Bell and Barbara M. Kennedy (eds.), *The Cybercultures Reader* (London, Routledge, 2000), pp. 291–324, http://faculty.georgetown.edu/irvinem/theory/Haraway-Cyborg-Manifesto-1.pdf, accessed on 24 October 2016.
3 Cf. Bruno Latour, *We Have Never Been Modern* (Cambridge, MA, Harvard University Press, 1993).
4 Bruno Latour, *Reassembling the Social: An Introduction to Actor-Network-Theory* (Oxford, Oxford University Press, 2005).

GESCHE JOOST

Although people tend to see humans and technology as polar opposites for understandable reasons, this isn't much use if we want to shed light on our networked world. We need to think in terms of new actors and new alliances that are based on the assumption that we are always online, that everything can be potentially networked with everything else, and that a significant part of human labour can be taken over by distributed computing. This stokes the fear of a loss of control: What will happen if we can no longer keep new technology at bay, or if it starts to operate autonomously and defy our instructions? Will the robot in its network become a legal entity with full responsibility for its actions? What powers will unshackled technology develop when it further evolves through an independent learning process, like the little kilobots that interact in a swarm?

Kilobots are small, autonomous bots that communicate with one another in swarms.[5] They can only perform a few actions on their own, as all they are able to do is orientate themselves in space. However, in a swarm they adapt their behaviour to one another, enabling them to assume different constellations. Each of them becomes a molecule within an organism that can change its shape.

The idea is that nanobots, the kilobots' miniaturised colleagues, are so small that they can be used as an autonomous army in the bloodstream to combat cancer cells. Although this may sound like science fiction at the moment, the first tests with multifunctional nanoparticles have in fact proved successful.[6]

5 Michael Rubenstein, Alejandro Cornejo, and Radhika Nagpal, "Programmable self-assembly in a thousand-robot swarm", *Science* (345, no. 6198, 15 August 2014), http://www.eecs.harvard.edu/ssr/publications/index.html, accessed on 24 October 2016.
6 Yuanpei Li et al., "A smart and versatile theranostic nanomedicine platform based on nanoporphyrin", *Nature Communications,* http://www.nature.com/articles/ncomms5712, accessed on 24 October 2016.

A swarm of kilobots can generate a huge variety of shapes. © Photo: Michael Rubenstein

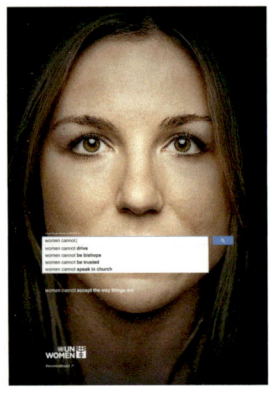

UN Women, *The Autocomplete Truth: "Cannot",* advertising campaign © Photo: Mermac Ogilvy & Mather Dubai

7 "A bot is considered to be a computer program that carries out repetitive tasks largely automatically, without depending on an interaction with a human user." From: https://de.wikipedia.org/wiki/Bot, accessed on 16 October 2016.
8 http://www.unwomen.org/en/news/stories/2013/10/women-should-ads, accessed on 16 October 2016.

110

UN Women, *The Autocomplete Truth: "Need",* advertising campaign © Photo: Mermac Ogilvy & Mather Dubai

Swarm robotics thus represents a very different kind of artificial intelligence that does not seek to replicate human intelligence. The individual bot[7] is relatively stupid – its intelligence arises from its interaction with the swarm and the autonomy this gives it. There is still a huge amount of research to be done on the various potential fields of application, but swarms of autonomous drones are one example of the new possibilities that beckon. However, this swarm intelligence also involves an increasing loss of control if the system is self-referential, learns from itself, and continues to evolve in this way. Self-learning systems like this are key in artificial-intelligence research, since they make it possible to keep developing the system based on the changing parameters of the environment. They are based on self-learning algorithms, which poses a crucial question: Who is responsible for these algorithms? Is it the programmer as their creator? The company that employs him or her? Or does the system evade responsibility because it has evolved autonomously – without human supervision? Take Google's intelligent search algorithm, for instance: the organisation UN Women attracted attention when it ran a sensational campaign on discrimination against women on the Web, in which it showed Google's auto-complete function for the search queries "women cannot" and "women need to".[8] The first listings in response to the query "women cannot" were "drive", "be bishops", "be trusted", "speak in church". In response to the query "women need to" came the results "be put in their place", "know their place", "be controlled", "be disciplined".

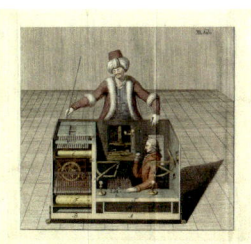

9 Nick Bostrom and Eliezer Yudkow-
sky, "The Ethics of Artificial Intelligence", in Keith Frankish and
William M. Ramsey (eds.), *Cambridge
Handbook of Artificial Intelligence*
(Cambridge University Press, 2014),
https://intelligence.org/files/Ethicsof
AI.pdf, accessed on 24 October 2016.
10 https://www.openai.com,
accessed on 24 October 2016.

Plate III from Joseph Friedrich Frei-
herr zu Racknitz, *Über den Schach-
spieler des Herrn von Kempelen und
dessen Nachbildung* (Leipzig and
Dresden, Joh. Gottl. Breitkopf, 1789).
Copperplate engraving © Humboldt
University of Berlin, University
Library; HZK

These results naturally raise the question of responsibility: Is Google as a company responsible for them, or is it the programmer of the algorithms, the crowd of users who entered the query, or even the system itself? Or does this self-learning system evade any responsibility because it merely represents what the mass of users have entered as a search query? This example shows how clichés and mass opinions are amplified by the Internet and how existing inequalities are consolidated. In this sense even an algorithm is not objective; rather, its action is reinforced by the emphasis of the search queries, thus enshrining world views without reference to an ethical or social substructure. A debate on the ethical bases for designing systems of artificial intelligence erupted some while ago in the developer community,[9] with initiatives like Open AI calling for radical openness and transparency with regard to AI systems, to be achieved via open source licensing, non-profit R & D, and security.[10]

MY COLLEAGUE, THE ROBOT

Robots are part of the interconnected environment and operate in networks. This gives rise to collaborations between humans and machines and between the machines themselves – what is called machine-to-machine communication. The vision of robots – which once came in the shape of automata but have been known since 1920 as robots, the name given them by Josef Čapek – has always fascinated the observer, while at the same time arousing fears of a loss of control. This vision pinpoints a primal human fear that runs right through our cultural history. The early examples of automata in the eighteenth century captured the imagination by virtue of their elaborate mechanisms, their precision, and the aura surrounding objects apparently coming alive. It transpired that one of the first major fakes was the Turk, the automaton chess player built by Wolfgang von Kempelen in 1769. A life-size figure sat in front of a chess board and responded to the moves of its opponent. It was supposed to compete with its human opponents as the first intelligent chess automaton, and indeed performed with flying colours – which should come as no surprise as it was in fact controlled by a diminutive person hidden in the automaton's torso. Until the early twentieth century the Turk continued to make appearances at carnivals and fairs and captivated the public as a symbol of the machine's triumph over the human mind.

11 Crowdworking refers to a new form of work in which a mass of users (the "crowd") take over tasks in the Internet for a limited period – these tasks are allocated via a platform. The collaboration between the workers and the companies behind the scheme is short-term and non-binding. **12** www.mturk.com, accessed on 24 October 2016. **13** The term gig economy refers to a current trend in which employment relationships typically cover limited periods of time and companies enter short-term project contracts with independent workers that are free of commitments – similar to a music band that goes from one "gig" to the next. **14** http://simonegiertz.com, accessed on 24 October 2016. **15** https://www.youtube.com/watch?v=g0TaYhjpOfo, accessed on 24 October 2016.

This automaton went down in history as the "Mechanical Turk" and inspired Amazon to start a crowdworking[11] platform with the same name.[12] The system is called "Artificial Intelligence", because it inverts the power dynamic between human and computer. It generates tasks that are offered on the platform and can be carried out by the "crowd" of users in exchange for minimal remuneration. A typical task is to transcribe data or compile product information for Amazon's e-commerce platform. Here the crowd of users supports the networked system by performing tasks that are difficult for artificial intelligence (AI) to fulfil or would simply be more expensive – perhaps a foretaste of the kind of new jobs that can be done by AI in the "gig economy" using cheaper crowd-workers.[13]

The story of the Turk only came to an end in 2016, when almost 250 years after the first automated chess player, Google's AlphaGo software managed to beat the Go world champion Lee Sedol, thus consigning the story of this fake to history. Newspapers ran the headline "Google beats human", heralding the victory as a key stepping stone on the path to AI dominance over humans. Having spent years developing a system capable of defeating the world champion in such a complex game as Go, the research world celebrated this as a breakthrough. In this instance the competition between human and machine was decided in favour of the machine, thus honouring the promise augured by the Turk all those years ago.

Simone Giertz paints a very different picture of intelligent automata.[14] With her self-built daily helpers the Swedish inventor and designer shows the limitations of machines, as they struggle to perform even simple gross motoric actions like chopping carrots, teeth-cleaning, or shaking breakfast cereal out of a packet.

This is precisely the problem with today's artificial intelligence systems: they are very good at tackling highly specialised jobs, while they fail miserably at everyday tasks. Motor coordination of the kind required in ball games is a highly complex chain of perception and reaction that machines have as yet barely managed to get to grips with. There are legendary videos of robot runners who simply fall down or trip over themselves when the terrain is the slightest bit uneven.[15]

16 Carl Benedikt Frey and Michael A. Osborne, *"The Future of Employment: How Susceptible Are Jobs to Computerisation?"* (2013), http://www.oxfordmartin.ox.ac.uk/downloads/academic/The_Future_of_Employment.pdf, accessed on 24 October 2016.

17 Erik Brynjolfsson, Andrew McAfee, *Race against the Machine: How the Digital Revolution Is Accelerating Innovation, Driving Productivity, and Irreversibly Transforming Employment and the Economy* (Lexington, MA, Digital Frontier Press, 2011).

18 Melanie Arntz, Terry Gregory, and Ulrich Zierahn, "The Risk of Automation for Jobs in *OECD* Countries: A Comparative Analysis", *OECD Social, Employment and Migration Working Papers* 189 (Paris, 2016), http://www.zew.de/publikationen/the-risk-of-automation-for-jobs-in-oecd-countries-a-comparative-analysis/, accessed on 24 October 2016.

What role will these autonomic machines play in our lives? Will it be that of Mary Shelley's wretched man-made Frankenstein, who is destroyed by his social isolation and becomes a killer? Or will it be more like that of Eva from Metropolis, a mechanical diva as a sensual female figure, an object of fascination? Robots have long since become a reality in our daily work routines, operating in highly specialised industries where they take over specific tasks previously handled by people. A wide-ranging debate was initiated by Carl Frey and Michael Osborne[16] with their prognosis of a massive loss of jobs on the US market: in their view 47 percent of jobs across all sectors were threatened by digitisation. In their book *Race against the Machine,* Erik Brynjolfsson and Andrew McAfee[17] argue that, just as the field of knowledge work will be affected by artificial intelligence, industry and manual trades will be hit by automation and robotics. Even if more recent studies take as their starting point far less shocking figures[18] (in 2016 the OECD referred to 12 percent of professional groups facing a potential threat), they still raise basic questions about our model of work if jobs become a commodity in short supply. Robotic systems have many advantages for networked production: they are efficient, they can perform specific tasks and routine activities with a high degree of precision, they can work around the clock, and they are not unionised. Used on a mass scale, however, they will shake up our social system, ultimately posing the question, What will happen if we run out of work? Ideas for a "machine tax" to be levied on companies using robots are already being forged by politicians, and calls for an unconditional basic wage as a means to provide social security for those who have become unemployed are becoming louder. The sharing economy of recent years is mutating into the gig economy, in which workers move from project to project as freelancers (like bands from one music gig to another) without having a fixed contract or being tied to one company. Crowdworking is the current buzzword, and the variety of jobs undertaken is becoming the modular assembly kit of working life. This might be a recipe for a new flexibility at work – the ability to be self-determined and combine and customise activities, side by side with new robotic colleagues who relieve us of the need to perform burdensome and repetitive tasks – if it weren't for the dangers of precarious employment and the loss of social security and the sense of meaning inherent to many professions.

113

Yet many also see new opportunities in digitisation: decentralised production in fab labs, for instance, lets anyone become an independent maker, using new technologies like rapid prototyping and, in particular, 3D printing. The open source ecosystem gives everyone free access to knowledge about procedures and programs as long as they are shared further under a Creative Commons licence. In this way small batches can be manufactured quickly and inexpensively, production returns to the cities, and access to the technology and know-how infrastructure becomes the key factor in self-determined work. Fab labs were conceived by their founder, Neil Gershenfeld, as germ cells for these new forms of production; the task was to create an infrastructure in which (almost) everything can be cheaply produced ("How to make almost anything").[19] In recent years this concept has proven especially successful in the emerging markets of India and Africa, resulting in the emergence of maker spaces and innovation hubs, which offer people local access to digital production technologies with which to develop their own business ideas and products. Open source is becoming the key concept in this movement: open source hardware and software provide access to the knowledge of the global community, outside the realm of companies and research institutes. It can also be seen as an ethical concept that views access to this knowledge as a form of empowerment through technology, as a basic right that seeks to harness the unique opportunity offered by the Internet to potentially allow anyone around the world to have access to knowledge. One symbol of the maker movement is the open-source 3D printer *RepRap,*[20] which can reproduce itself. Like a perpetuum mobile – only in this case with an electricity supply – this printer can print itself and so expedite an infinite chain of reproduction. The blueprint is available to anyone on the Internet at low cost and is the project of a community that has set itself one goal: to make rapid prototyping technologies available to everyone and to further develop open source – into a decentralised and globally networked phenomenon existing beyond established corporate structures.

Thus the maker movement espouses the high-minded aspiration to use technology to improve the world – the keywords here are empowerment, as the potential to be personally involved in the design of new technologies from anywhere on the planet; sharing, as the possibility of distributing knowledge worldwide; and participation, as the opportunity to take part in the networked society outside the established structures of companies and institutions.

The fact that the digital divide is nonetheless becoming deeper around the world – that even in the industrialised nations digital education is not sufficient to really allow participation in the digital society or a proper understanding of the "black box" of digital technology – is quite another matter. The maker movement remains the preserve of the globally networked, digital elite, even if Chris Anderson[21] speaks of the makers prompting the next digital revolution, while Jeremy Rifkin[22] predicts that they will spell the end of capitalism. At present the situation is still unsatisfactory – the digital divide also sees society drifting apart as many people feel themselves left behind and fear the end of the right to privacy, technology-induced unemployment, and a worsening of social inequality as a consequence of insufficient education.[23]

19 Neil Gershenfeld, "How to Make Almost Anything: The Digital Fabrication Revolution", *Foreign Affairs 91,* (no. 6, November / December 2012), http://cba.mit.edu/ docs/papers/12.09.FA.pdf, accessed on 24 October 2016.
20 http://reprap.org, accessed on 24 October 2016.
21 Chris Anderson, *Makers: The New Industrial Revolution* (London, Random House Business, 2012).
22 Jeremy Rifkin, *The Zero Marginal Cost Society: The Internet of Things, the Collaborative Commons, and the Eclipse of Capitalism* (New York, NY, Macmillan, 2015).
23 Initiative D21, *D21-Digital-Index 2015: Die Gesellschaft in der digitalen Transformation,* http://www.initiatived21.de/portfolio/d21-digital-index-2015/, accessed on 24 October 2016.

THE DATA SUPPLY

"Data is the new oil" is one of the emerging platitudes of the twenty-first century. Yet it is still true that the possibilities offered by data analysis and data trading under the banner of Big Data have changed our society. The connection of everything to the Internet is becoming the norm so that data is generated everywhere. Equipping our pets with RFID chips so that only the right cat can get into our home is now a standard service at the vet's. Many people shudder at the thought of having a chip like this implanted in their own bodies – but more because the needle for the tiny chip is still fairly large and the procedure somewhat brutal as a result. We are no longer surprised when our artificial hip joints are checked for wear and tear using a chip or our pacemaker is monitored via the Internet. So far there has been little discussion of the underlying questions about our privacy that this state of being "always online" raises. Who owns the data that is generated in this way – I as patient, the doctor who uses it for a study, or the insurance company that wants to track the efficacy of its investment? A struggle has broken out over the sovereignty of data; and this is just the beginning, as the potential inherent in the utilisation of our data starts to dawn on us. As yet, only a minimal part of this data is analysed because it is still not clear what it can be used for. Intelligent services on databases are still in the pipeline – such as the repair service that gets in touch shortly before the dishwasher gives up the ghost. This scenario, which is called "predictive maintenance", is being worked on by the R & D departments of large companies. The sci-fi story *Minority Report* anticipated this technology in another field – as a means to fight crime before an offence is actually committed. And indeed Big Data can be used to calculate the probability that crimes like assault or mugging will take place at a particular time of day in a particular neighbourhood – not definitely but with a fair degree of likelihood. Does this probability imply responsibility – as it does in *Minority Report?* Does potentiality assign blame?

Now, rather late in the day, we are asking who actually has control in these kinds of merged worlds. Has technology become autonomous? Does it threaten to defeat us, as it did the Go world champion in 2016? There is no longer any need for direct human control to optimise either the logistics in the Port of Hamburg (this job has been taken over by intelligent algorithms and a network of sensors) or the mobility of a city (intelligent traffic management, resource-friendly vehicle utilisation, and flexible congestion alerts can be controlled more efficiently via Big Data analysis). We have delegated these processes to technologies, but they are still underpinned by human choices as to how they are to be designed.

The networked world is becoming more complex as decisions are taken in the network of actors. Clear divisions between human and machine are a thing of the past. We will be confronted with a reformulation of fundamental questions: Who bears responsibility in the network of actors? Who owns my data and where does my private sphere begin?
How can one live as a self-determined individual in a networked society and retain digital control? We might want to discuss these questions with our new robot colleagues.

Gesche Joost (born in 1974 in Kiel, Germany) studied design in Cologne, writing her dissertation on rhetoric. She was Professor for Interactive Design and Media at the Technical University, Berlin, until 2010. Joost was the founding Board Director of the German Society for Design Theory and Research. As a member of several advisory committees for the German government, she has been highly influential in shaping the concept of Industry 4.0. Since 2005, she has run the Design Research Lab at the Berlin University of the Arts. The Lab concentrates on interdisciplinary design research projects at the interface of technological innovation and human needs. Joost lives and works in Berlin.

DO YOU
WANT TO
BECOME
A PRO-
DUCER
YOUR-
SELF?

COULD A ROBOT DO YOUR JOB?

Manufacturing #10a, Cankun Factory, Xiamen City, China, 2005

Manufacturing #10b, Cankun Factory, Xiamen City, China, 2005

EDWARD BURTYNSKY – *MANUFACTURING #10AB, CANKUN FACTORY, XIAMEN CITY, CHINA*

With his photo story about contemporary China – for which he visited the country's most important shipyards, industrial plants, coal-mining areas, and cities – Edward Burtynsky allows us to see and feel how global mass consumption is organised. In this photograph of a 450-metre assembly hall of the Cankun Factory, the world's second-largest manufacturer of coffee machines with 23,000 employees at the time the photograph was taken, we see hundreds of workers. They all wear the same clothing, occupy similar workspaces, and perform nearly identical operations. Burtynsky's photographs wordlessly show how people become machines to a certain extent even before the machines replace them. TT

118

The installation consists of a robotic scribe which constantly writes manifestos, throwing the paper on which each is written at the visitors as soon as it is complete. Every manifesto consists of eight statements, which the robot generates autonomously by selecting terms out of its internal information pools (on the subject areas of art, philosophy and technology) and associating them within a syntactical framework. In other words, rather than mass-producing copies of one uniform manifesto, the machine mass-produces unique texts with individual messages, each one signed with a serial number.

Despite their uniqueness, however, the manifestos are random, automatically generated, and devoid of intentional meaning. *manifest* is supported by Kuka. AR

ROBOTLAB – *MANIFEST*

robotlab. *manifest,* 2008. Industrial robot installation, 180 × 240 × 120 cm; thanks to: KUKA and ZKM © robotlab

manifest, 2008

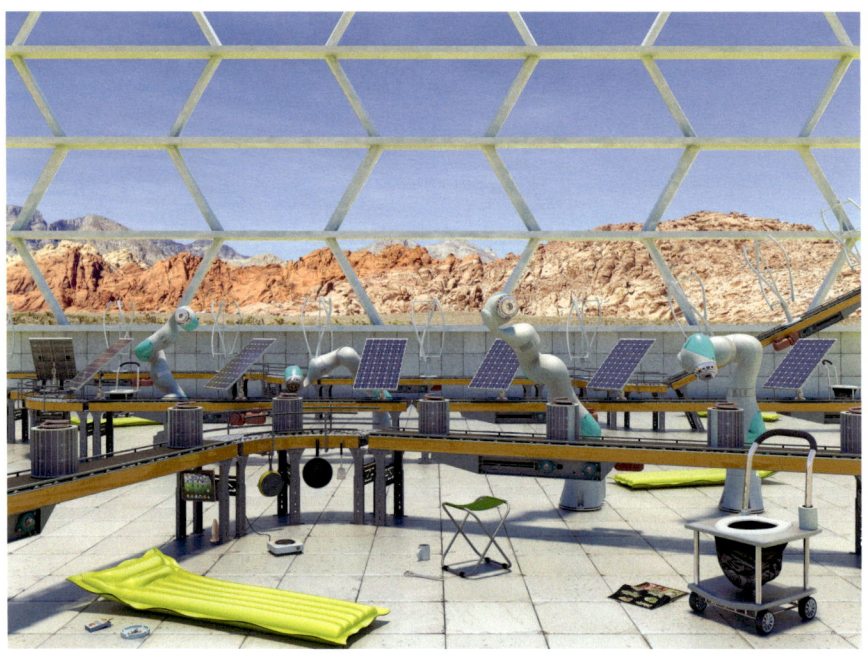

Going Green, 2016

The artist Shawn Maximo creates visions of possible future living spaces by mixing various concepts of spatial utilisation, which is why his photographs seem at the same time both strange and familiar. In *Going Green,* we see a production hall with autonomous robots working on a manufacturing line producing "green technology". In the foreground of this scene devoid of humans we see a scattering of what looks like camping equipment: an air mattress, a folding chair, a camping stove. It begs the question of the role of humans in this post-industrial scenario. Have they already been made obsolete in the work force? Or are we dealing with a future concept of spatial use that does not yet strike us as possible or even rational? TT

Shawn Maximo. *Going Green*, 2016.
Vinyl print, 356 × 491 cm © Shawn
Maximo

SHAWN MAXIMO – *GOING GREEN*

Based on two current studies on the topic of automation in the workplace, the BBC's interactive website allows visitors to calculate the risk that their jobs will be taken over by robots or computers in the near future. Predictably, jobs relying on empathy or social intelligence, such as work in the nursing or creative fields, are not under immediate threat. Manual and elementary administrative jobs, however, face a much greater risk. Here, as is often the case, automation – in the form of more or less likeable humanoid robots with deep-set eye slits – is described as if it were something akin to a force of nature rather than a process that could be addressed and possibly countered at a political level. TT

BBC NEWS – *WILL A ROBOT TAKE YOUR JOB?*

Will a robot take your job?, 2015: the website ...

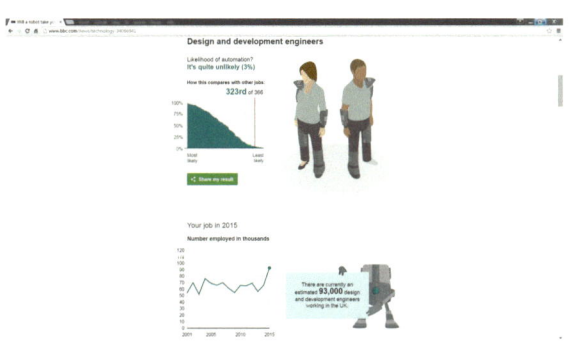

... calculates your risk of being replaced by automation.

BBC. *Will a robot take your job?*, 2015. Interactive website; team: Nassos Stylianou, Tom Nurse, Gerry Fletcher, Aidan Fewster, Richard Bangay, and John Walton; screenshot © BBC, courtesy Getty Images / BBC Motion Gallery

The animated film *The Last Job on Earth* describes a scenario in which human labour has been eliminated: everything from human hygiene and housework to mobility, medical care, news, and shopping is now the task of computers or robots. But protagonist Alice still has a job – for now. The viewer accompanies her as she goes about her morning and makes her way to the office. What is revealed is not only the "brave new world", but also its price: growing inequality, slums, people left behind by the pace of progress, and the problem of providing for a population condemned to idleness. The scenario raises the question of whether a society without work – all technical progress aside – is at all desirable. TT

MOTH COLLECTIVE AND BOX OF TOYS AUDIO LTD. – *THE LAST JOB ON EARTH: IMAGINING A FULLY AUTOMATED WORLD*

Machines don't always work in the future either.

Moth Collective and Box of Toys
Audio Ltd. for *The Guardian. The
Last Job on Earth: Imagining a Fully
Automated World,* 2016. Video,
2 min 53 sec © 2016 Guardian News
& Media Ltd.

The Last Job on Earth, 2016

... everything looks pretty great, but only superficially.

JULIUS BREITENSTEIN – *THE UNPAID INTERN*

124

For his 2016 Graduate show presentation at Central Saint Martins in London, Julius Breitenstein designed *The Unpaid Intern*. With more discussion about (design) jobs losing out to new, computer-aided and autonomous technology than ever before, Breitenstein's contribution is well-timed. Instead of minimising the designer's role, *The Unpaid Intern* aims to do the opposite – support designers by using genetic algorithms as tools for improving the design process. The square-shaped controller Breitenstein created employs a straightforward CAD user interface, is parameter-based and allows users to experiment a great deal in the early stage, save their work as presets, and combine things later. The device offers designers more efficiency and choices, but the real interns might be stuck getting the coffee. EP

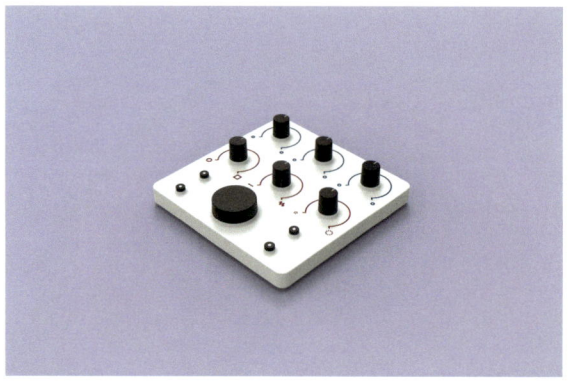

The Unpaid Intern, 2016: controller

Julius Breitenstein. *The Unpaid Intern,* 2016. Algorithm-based design software; controller, ca. 17 × 17 × 4.2 cm; 3D printed forms, various sizes © Julius Breitenstein

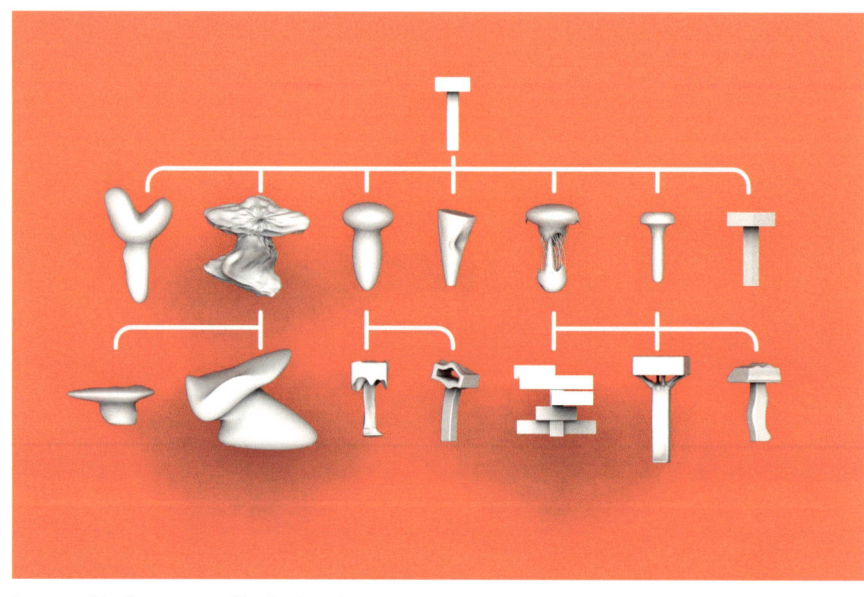

Diagram of the forms generated by the algorithm

YuMi, 2015

YuMi is a dual-arm robot designed to perform small-part assembly work required, for example, by the electronics industry. It has multi-functional, sensor-equipped "arms" and extremities with fourteen rotational axes. *YuMi* – whose name derives from the words "you" and "me" – is as dextrous as a human being, with arms and grips that mimic those of a human in both form and function. While traditional industrial robots are kept apart from humans by protective barriers – because they cannot sense when a human is nearby – *YuMi* is a collaborative robot, designed specifically to work safely with and around people. With simple programming, it is can perform any manual task and, just as the steam engine replaced human muscle, it will increasingly take care of repetitive assembly tasks allowing humans to focus on value-creating activities. TT

ABB Ltd. *YuMi,* 2015. Collaborative robot, total floor space: 39.9 × 49.6 cm, arm span: 50 cm, weight: 38 kg © ABB Ltd.

ABB ROBOTICS – *YUMI*

The scenario presented in the video *Teacher of Algorithms* assumes that our "smart" devices, such as those we use in the home, are actually not all that smart, since they cannot be truly effective helpers without first learning and observing our habits and adapting to them. But what about when the coffee machine, for example, begins to regularly make coffee at two o'clock in the morning after its owner worked all night? Wouldn't it be practical if we could take our devices to an algorithm coach who could professionally train them according to our needs, thereby turning them into devices that are truly smart? After all, new technologies always create new job profiles in their wake. TT

SIMONE REBAUDENGO, AUTOMATO.FARM – *TEACHER OF ALGORITHMS*

Simone Rebaudengo (automato.farm). *Teacher of Algorithms,* 2015. Mixed media installation and video, 5 min 28 sec; script, direction, editing: Simone Rebaudengo; script, sound: Daniel Prost; camera: Andrea Carlon © automato.farm

Teacher of Algorithms, 2015

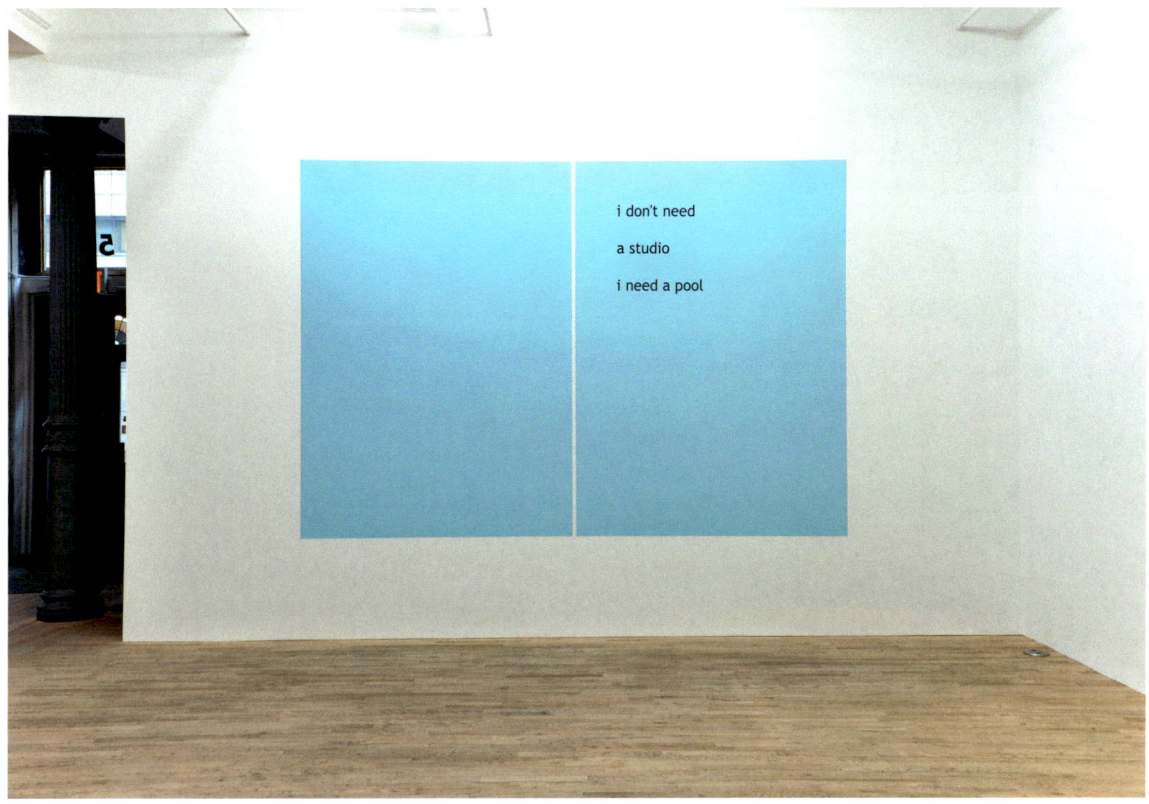

RR Haiku 154, 2015

RAFAËL ROZENDAAL – *RR HAIKU 154*

Rafaël Rozendaal. *RR Haiku 154*, 2015. Painting, paint and vinyl lettering, 250 × 167 cm © Rafaël Rozendaal

In 2013, the artist Rafaël Rozendaal began posting haiku on his various social media accounts that, along with a general commentary on society, often assess the nature of work today – from the banality of too much screen time: "clickidiclickclick, clickclickclickclickidiclick, clickclickidiclick" (*RR Haiku 68*), to our inability to step away from the screen: "never working, never, not working." (*RR Haiku 111*). His haiku have since found their way off the screen and into other media such as a book published in 2015 and various large-scale wall paintings. *RR Haiku 154* – "I don't need a studio, I need a pool" – was chosen for the exhibition *Hello, Robot.* because it playfully questions what we will do with all of our newfound free time when robots take over our jobs: Will something else keep us confined to our desk, or will we hightail it straight to the pool? EP

DO YOU WANT TO BECOME A PRODUCER YOURSELF?

128

COU

OBOT DO YOUR JOB?

Vogt + Weizenegger. *Sinterchair,*
2002. Chair, laser-sintered nylon,
76.2 × 44 × 50.5 cm © Vogt +
Weizenegger, photo: Jürgen Hans,
courtesy Vitra Design Museum

130

Sinterchair uses the sintering process – previously employed in the automobile and aviation industry to build prototypes – to manufacture custom-made, individually designed chairs. Customers are first given a questionnaire to find out their preferences, such as their favourite music, writers, and philosophers. Next a computer-generated drawing of a chair is created. Subsequently the selective laser sintering (SLS®) process is used to outline a chair in a block of nylon powder and the shape is cut out one layer at a time. The chair takes shape as the layers harden. A few hours later, the chair is removed from the nylon block and the customer can take it home. *Sinterchair* places the consumer at the centre of the design process while minimising distribution, storage, and model costs. AR

Sinterchair, 2002

VOGT + WEIZENEGGER
– *SINTERCHAIR*

TAL EREZ – *BANDE À PART*

Installation view

Bande à part, 2013: 3D printed car

Bande à part was created by Tal Erez in 2013 for the Gallery Z33 exhibition *Design Beyond Production* shown in Hasselt, Belgium, and at the Salone del Mobile in Milan. The critical design installation illustrates the political, social, and ethical repercussions of the designer-consumer-manufacturer paradigm motivated by advancing DYI technology such as 3D printing. The installation has been scaled down for *Hello, Robot.,* yet it maintains its original three features: command center, control screen, and assembly line, all of them with a designer at the helm. In this way, a distributed factory is simulated, in which individual households print out goods under the surveillance of the designer. As the designer's role shifts to the one of an overarching manufacturer, his or her responsibilities multiply accordingly, thus going dramatically unregulated. EP

Tal Erez. *Bande à part,* 2013. Mixed
media installation with ten 3D
printed cars, PLA, 12 × 6 × 7 cm each
© Tal Erez

Command centre

The interactive installation presented at CeBit 2015 by the designer duo Kram and Weisshaar
sought to demonstrate how the intelligent production systems of the future can enable users
to design and manufacture their own products with the aid of industrial technology. On a touch-
screen, visitors could create their own designs, which were then cut from a block of foam
material by a hot wire-cutting tool. The resulting objects were then shipped free of charge to
their creators' homes all over the globe. Many of Clemens Weisshaar's and Reed Kram's earlier
projects also focused on the opportunities for individual manufacturing provided by automated industrial production processes. TT

KRAM / WEISSHAAR – *ROBOCHOP*

Robochop, 2015

DIRK VANDER KOOIJ
– ENDLESS FLOW ROCKING CHAIR

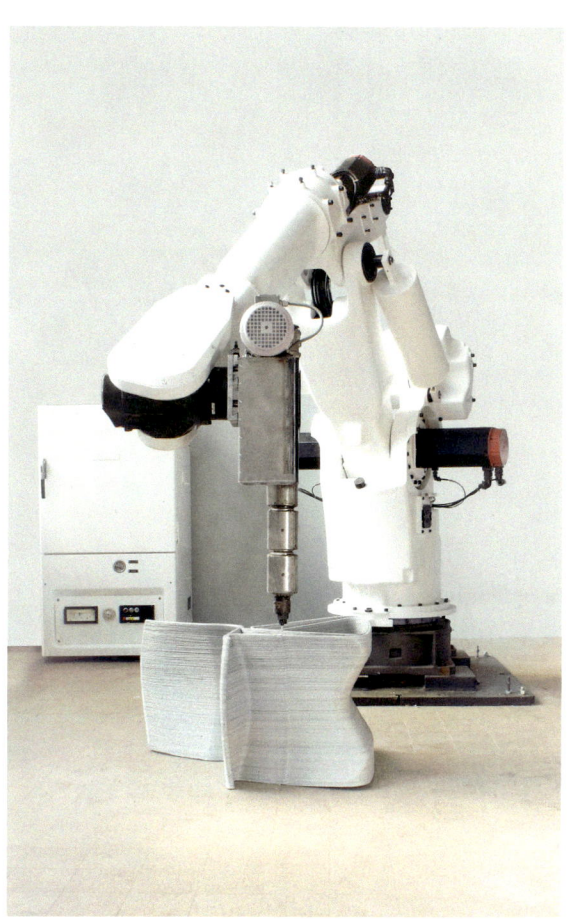

Production

Endless Flow Rocking Chair, 2011

For the *Endless* collection, Dirk Vander Kooij created the world's first robot capable of 3D-printing furniture from 100-percent recycled material. Melting down the finely chopped innards of old refrigerators, the revamped industrial robot extrudes one endless plastic string and bends it back and forth to form the final object. This kind of low-resolution 3D printing produces a chair in just three hours. The technology also enables the designer to modify a model after a piece of furniture is produced   a bonus that the traditional injection moulding process cannot offer. The machine can be programmed to build furniture of any shape and size. The *Endless* chair won the Dutch Design Award in 2011. AR

Dirk Vander Kooij. *Endless Flow Rocking Chair,* 2011. Chair, recycled plastic, 82 × 42 × 68 cm © Dirk Vander Kooij, photo (l.): Jürgen Hans, courtesy Vitra Design Museum, photo (r.): Studio Dirk Vander Kooij

GRAMAZIO KOHLER RESEARCH – *ROCK PRINT*

The robot lays out the string.

Gramazio Kohler Research. *Rock Print,*
2015. Architectural installation with
industrial robot © Gramazio Kohler
Research, ETH Zurich

Rock Print, 2015: Installation view at the Chicago Biennale of Architecture

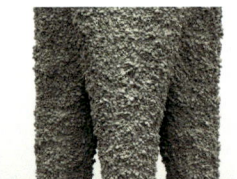

Detail view

The architectural form – a sort of four-legged column – of Swiss architects
Fabio Gramazio and Matthias Kohler (and their eponymous research
institute at the ETH in Zurich) uses a new technology in which coarse
particles, in this instance foam glass aggregate, are combined to create
stable objects. In this technique, robotic arms lay out a recycled textile string
according to a particular algorithm, each layer of which alternates with
a layer of stones spread manually on top of the string. Layer by layer, the
structure takes shape within a wooden form, until the stones held by
the string structure can stand on their own. *Rock Print* thus functions in
a manner similar to 3D printing, except that it uses a mechanical rather
than a chemical sealer. The mechanical sealer also has the advantage of
being reversible by the simple expedient of pulling out the string. TT

Individuality, intricacy, and beauty are words that have only recently become linked to manufacturing processes utilising industrial robots. Hyunchul Kwon, Amreen Kaleel, Xiaolin Li (Team CurVoxels), a team of three students at the UCL Bartlett School of Architecture, developed software within the framework of their graduate studies programme to produce a filigree cantilever chair based on the form of Danish designer Vernon Panton's legendary Panton Chair. However, while both designs target mass production, the CurVoxels model presupposes *individualised* mass production based on variable algorithms. After the team plugs in a computed algorithm, the robot creates the chair's intricate form via one continuous motion, allowing the extruded plastic to cool down on the way. The resulting details are nothing short of spectacular. EP

CURVOXELS – *3D PRINTED CANTILEVER CHAIR*

CurVoxels, Research Cluster 4, UCL the Bartlett School of Architecture. *3D printed cantilever chair,* 2015. ABS plastic, 45 × 50 × 85 cm; video, 3 min 27 sec; team: (CurVoxels) Hyunchul Kwon, Amreen Kaleel, Xiaolin Li; tutors: Gilles Retsin, Manuel Jiménez Garcia; technical support: Vicente Soler Senent, William Bondin © CurVoxels 2017, photo: Sin Bozkurt, CurVoxels

3D Printed Cantilever Chair, 2015

Space Exploration Architecture /
Clouds Architecture Office. *Mars Ice
House,* 2015. Architectural model,
3D printed resin, white corian, 26.5 ×
66 × 66 cm © Clouds AO / SEArch

After NASA recently discovered evidence for the existence of water on Mars, the possibility of living on the planet appears to have moved out of the realm of science fiction and become a real possibility. A competition sponsored by the American space agency NASA, in which participants used 3D printing technology and materials naturally occurring on Mars to construct living quarters for four astronauts, was won by the designer team made up of Clouds AO and SEArch. Their twin shell structure "printed" from ice, in which plants supply oxygen for its inhabitants, subsequently protects its occupants from the planet's inhospitable climate and temperatures that can reach -150°C. At the same time, the transparent nature of the building materials and large windows allow contact with the outside world. LH

136

Mars Ice House, 2015: Exterior view

SPACE EXPLORATION ARCHITECTURE / CLOUDS ARCHITECTURE OFFICE – *MARS ICE HOUSE*

Details from the interior

... built by a robot.

JORIS LAARMAN – *MX3D BRIDGE*

The project *MX3D Bridge* by the Dutch designer Joris Laarman opens up entirely new dimensions in the field of 3D printing. Bridge construction normally requires a number of different techniques and specialists, but Laarman's pedestrian bridge is to be printed in steel across a canal in Amsterdam by means of six-axis industrial robots, meaning that the project will rely solely upon a single, universally applicable, fully robotic manufacturing technique. Although 3D printing is usually associated with the manufacturing of smaller, more delicate objects, with this project the technology has made its breakthrough into the wider world of architecture and construction. TT

Production

Joris Laarman for MX3D. *MX3D Bridge*, 2015 (ongoing). 3D printed bridge, rendering © Joris Laarman Lab

MX3D Bridge, 2015 (ongoing)

FRIEND AND HELPER

We are already reliant on intelligent devices in our day-to-day lives. We trust our smart assistants to navigate us through foreign cities, to remind us of our anniversaries, and to provide us with information on every subject imaginable. They check our pulses and even call for help in an emergency. All these things have made our lives more comfortable and in some cases have saved them. But anyone who has had to make do without their smartphone for even one (working) day knows how dependent we have become on their intelligent help, and how helpless we are when they turn out to be nowhere near as smart as promised.

Our relationship with objects and how best to design it has preoccupied designers for decades. Whenever consumers have the choice between a variety of similar or identical products, the unique selling point becomes the sensory or emotional qualities of the objects. This is all the more true of intelligent objects that communicate and interact with us and give us the feeling that they can feel – because what counts then is not their shape or tactile properties, but how they succeed in making us reliant upon them. If roboticists have increasingly started talking about "humanised" machines, what they mean is that these machines should feel like old friends: helpful and obliging, perhaps a little over eager, and maybe even neurotic or manipulative. Once we are entangled in these relationships, their loss hits us all the harder. What happens if and when the beloved thing is gone forever?

The robots who look after us, who nourish and care for us, who make sure we are okay, are everywhere in society. We have yet to entrust our children to robotic nannies, but there are already a host of robots designed to be playmates, teachers, and chaperones, rolled into one. The wide-scale use of robots in geriatric care is the subject of serious discussion – not only in Japan, where people are traditionally more open towards intelligent machines, but also in the West. There are good reasons for this, since robots are already very successfully deployed in caring for and treating patients with dementia and Alzheimer's disease. Younger adults don't want to miss out on being cared for by robots either, and they don't have to – whether while shopping for jeans, having the shopping delivered, or taking off their new jeans for a bit of casual sex.

BY BRUCE STERLING

MY ELEGANT ROBOT FREE- DOM

I was born in 2016 in a mansion in Hannover. Our house was equipped with state-of-the-art home automation. We had all the finest gadgets that Hannover's CeBIT robot conference had to offer.

It took me a while to understand all this, but my rich family's home technology was amazingly bad. Their robots were rude, crude, and terrible. Now that I'm twenty-five years old, and we're in the robot Golden Age of the 2040s, I've achieved some perspective in these matters.

For instance, I now know that I never got proper human attention from my own parents. They were industrial robot entrepreneurs, always preoccupied with their labour-saving businesses. So they imagined that their home machinery could raise their child for them.

So, as I crawled the floor in my smart diapers, both my little wrists had medical scanner tags. My baby crib was saturated with twenty-four-hour video and motion trackers. My nursery had ten varieties of intrusion and environmental safety alarms.

When I grew a bit older, I found that the house tracked all my childish movements. My neighbourhood was a gated community. My smart city had security cams at every traffic light. Surveillance drones buzzed constantly overhead.

As soon as I was tall and strong enough, I rebelled against this stifling conformity. I escaped that grand, unhappy house and I fled for Munich to find a new life of freedom!

The Schwabing district is a traditional cultural refuge for us Germans, especially those among us who search for a spiritually higher, more ideal way of life. That was my quest. I had become a modern young woman of the 2040s. I was determined to participate in the moral struggles of my generation.

I was welcomed in the leafy and pleasant Schwabing cafes, where we, the youth rebels of an old-fashioned digital order, threw our mobiles into buckets of sand so that we could speak directly to one another.

Why? we all demanded. Why did we have so many algorithms, so many cameras, microphones, trackers, and actuators? Why did we have so little tenderness, care, and genuine physical and spiritual authenticity? Were we all doomed?

Our world was saturated with machine intelligence (a tangled web of fibre-optics, towers, and satellites, from pole to pole). So why couldn't I, Franziska von K., a young German woman with cultural aspirations, find one private space, with a door that locked against an intrusive world – one precious space where I could truly become myself? Was a room of my own too much for a woman's aspiration?

Schwabing was seething with a generation of young people who shared my moral crisis. We all deeply resented being raised by bad software and a torrent of social-media prompts. Thousands of crass and commercial home-management apps and bots had been designed by, and for, the use of grown-ups. All of that effort had been aimed directly against the emotional interests of us, the youth of tomorrow.

We realised that our parents' gadgetry had always been ruthlessly exploitative. Their web technology was hostile to civilised values, it was brutal, rude, and repressive, it was market-surveillance rubbish! The harsh life of our parents' electronic frontier had left scars on our souls.

I spent years of emotional tumult in Schwabing, raising our generational consciousness, and consoling the wounded among us, especially the sensitive, disturbed young men who had spent far too much time on computer games. I wasn't quite sure what I wanted to achieve with my own life, but I seemed to have a talent for the personal touch.

Eventually, I realised that we had to take action. It wasn't enough to lament together about our squalid zeitgeist and our twenty-first-century weltanschauung. We had to organise and fight to change our world for the better.

But how to do it? I found an answer – by reinventing the tools that had been used against us. Robots, mostly: because robots are "actuated, mobile, programmable mechanisms with some degree of autonomy".

Could we reinvent robots entirely? Could we chic Munich rebels create a new, advanced breed of robot devices that nobody had ever expected?

In Schwabing, I joined the nascent social movement for a nouveau-robotics lifestyle. My new friends were a loose cluster of programmers, designers, mechanics, architects, interior designers, and artisanal furniture makers. We were like our spiritual ancestors, Munich's brilliant Jugendstil movement, but we were even more cosmic!

Our goal was nothing less than the revival of European domestic life! We would no longer suffer from intrusive Internet of Things social software, mostly designed, badly, in California. Ours would be a liberated, elegant existence, life re-imagined as a total work of soft robotics.

We indignantly rejected Europe's traditional robot heritage (which, as everybody knows, dates back to Karel Čapek in Prague in the 1920s). We dismissed that older paradigm of robots as humanoid metal monsters.

We all desired "soft robots". Why?

Because we realised that the authentic need in robotic home design was a robot technology that was inherently home-like. Instead of roboticising the home, we aimed to invert the established industry and fully domesticate the robots.

The older generation's robots were all steely, masculine, cold, heavy, warlike, drone-like, and oppressive. Our generation would favour entirely new robots that were dainty, fluid, auxetic, lightweight, and totally civilised!

Our nouveau-robotic devices, deployed in domestic contexts, would have the soft, velvety, yielding character of beds, couches, pillows, cushions, carpets, and drapes. These were entirely new product categories of robots. We called them Blobs, Tentacles, Fabrics, and Lights.

Some of our household soft robots were so humble and cheap that they were entirely disposable, like tissue paper. Others were flat, compact, and flexible devices, like origami figures. Still other soft robots advanced and retreated, swelling and shrinking, like silken hot-air balloons.

The true key to our enlightenment was intelligent light. Robot vision would be built directly into the home lighting systems. Robot mapping, radar, and lidar all came directly from the intelligent track-lights mounted in the walls, floors, and ceilings. So light and robot vision would always be one and the same thing.

This important advance meant that the entire body of house became the "robot" that we lived within, while every machine inside the house was just a peripheral device, a replaceable component. This profound new paradigm enabled a soft robot liberation.

The technologies required to make robots softer were already at hand. All that was needed was the daring and the will to think and live differently.

Our parents had built old-fashioned, stand-alone, metal robots: stiff, precise, exact, and hopelessly stuffy and boring. Our new generation created bag-shaped, elegant cloud robots which were free to wobble and wiggle, slither and flow!

Our Robot Blobs were like velvet sandbags. They had no rough hinges, no pinch-points, and no fragile moving parts to break. Blobs posed no dangers to children. And Blobs had no spyware built in, because Blobs were almost nothing but flexible nanocarbon skin.

A soft Blob that thinned and stretched out became a Tentacle. The Tentacle was a soft, gripping robot manipulator, always tender and gentle. This lightweight, translucent, ultraflexible Tentacle was ideal for domestic tasks such as silently cleaning dishes or arranging fresh flowers. Our polite and elegant Tentacles were certainly a leap beyond those ugly metal gripper-arms that welded cars on dirty assembly lines.

As for the waving smart Fabrics and the glowing smart Lights, they were complementary; they were like media screens and augmented projectors.

These new potentials deeply excited me. I felt that I had found my cause, and I advanced in taste and understanding every day.

I charmed my way into all the garages, labs, and ateliers of the Munich robot design scene. Carefully, I ran my own hands over the self-tying, tuneable, translucent curtains. I inspected plastic caterpillars, butterfly drones, and self-propelled prototype barbecue kitchens. I personally tested the satin smart blankets and the tender, ultra-sensitive beds.

I was not just buying these soft-robot products (although I certainly did). I was acculturating them. I was never going to be a standard technician, designer, or inventor, but I was a modern woman spiritually moved by a profound desire to live authentically.

That meant that I could inspire people. I could personify a situation. All these lonely inventors, these obsessed technicians in their cluttered garages, they needed more from their lives than investor cash and a niche in a furniture market.

These modern artisans needed tender care and kindness. They needed undivided personal attention. They needed a sympathetic listener, a full-hearted and forgiving woman who could be trusted with their innermost sorrows.

They needed all those immaterial yet vital acts of spiritual nourishment that artists always need. When I realised that I could liberate myself by giving those precious gifts to other people – that's when I suddenly became rather famous. Because I was their muse figure. I had become "Soft Franziska", the It-girl that everybody in Schwabing talked about.

I left my sweet little penthouse and I relocated into a big downtown loft, where I could entertain the people of my scene. Drawing on my growing skills in interior design, I furnished my new loft exclusively with Blobs, Tentacles, augmented Lights, and interactive Fabrics. My house was a showplace of soft robotic possibility.

As the hostess of this lively, colourful, gently pulsating gesamtkunstwerk, I gathered furniture creatives every Wednesday night. All my soirees featured white wine, pretzels, and those tasty white Munich sausages.

My avant-garde loft became a locally famous meeting space, where the soft-robot influencers, thought-leaders, and their chosen guests could all gather to confront their cultural issues, without any fear that bankers would suddenly pounce on them.

It's never easy for a woman to console a large group of cantankerous men without marrying even one of them, but I had some help in these matters, from the worldly wisdom from my grandfather.

My dear old grandfather was a 1968er. So, even though he was ninety-one years old and well along into senility, he had always been the member of my family who understood me best.

My grandfather pointed out that my lavish entertaining would soon cause my trust fund to run out. Grandpa warned me that I would never thrive as a glamorous design muse and famous Munich socialite unless I found some kind of business model that would support me.

I had to respond to this moral pressure from the one who had always loved me best. So, I surprised my grandfather with a new retirement apartment.

My creative soft robotic friends and I carefully picked and chose the objects, functions, and services that were ideal for robotic eldercare. Then we had a big house-warming party and we installed Grandfather in his new soft-robot environment. (Of course he paid for all of that, but that was the cost for us accepting his good advice.)

Now my grandpa had a pneumatic, bubble-packed support bed, a tentacular massage chair, plus two fully autonomous sink and stove robots that would fetch his food from groceries and prepare all his meals for him. He never lost his keys or glasses anymore, because he had gigantic paper display screens for all his reminder messages.

I also gave my grandpa a single giant red STOP button that would turn off all robot services. One punch of this existential-freedom button would plunge his room into complete, inert, total, entirely untechnical darkness. Not one whisper of light, sound, data, or even electrical voltage. He was back to a hippie state of nature.

Grandpa used this big red STOP button quite a lot. He used to punch that button, light wax candles in the dark, take LSD, and then read Goethe.

Grandpa was especially interested in Goethe's famous colour theory. Grandpa would ingest his favourite 1968-style mind drugs, read his classic Goethe, become entirely fascinated, and then suddenly shout in outrage that Goethe's romantic concepts were completely absurd and crazy. He would rant about psychedelic colours until he collapsed.

On waking, Grandpa would have forgotten everything about it. He would cheerfully start reading Goethe again from the first page.

I'm relating this apparent digression – not to tell scandals against my poor ex-hippie grandfather – but to point out that the soft-robot room which I'd given him was ideal for this form of behaviour. I'd never seen my dear grandpa so engrossed and fulfilled as he was in his final days. Soft robotics had allowed him to lose his sanity, but never his dignity or his independence.

My grandfather peacefully passed away in the tender grip of his Tentacles. Heeding his good advice about business, I took Grandpa's bequest to me and I built two new soft-robot apartment rooms.

I quickly rented out all three of the rooms to eager tenants. I had become a Munich landlady.

My tenants were a mixed lot, since Schwabing is such a powerfully attractive neighbourhood. Iggy and Ina were a pair of bisexual artists; they were working on a huge, mysterious, interactive entertainment project. This scheme involved vast colonies of small Blob-robots, which could latch together like living cells to form a throbbing land-art monument. This thundering colossus of robot Blobs would roll majestically through the desert and then set fire to itself in public.

Iggy was a noted expert on humanoid robot gripper hands. She built huge robot hands that she sat in and slept in, and printed hundreds of tiny ones like busy little mouse paws. Ina was the intellectual of the pair. Ina used to travel everywhere with her purse full of flat, stick-on eyeballs. These robot eyeballs were networked cloud cameras, but Ina prized them for their emoji characteristics. We all loved the way her graffiti eyeballs would examine the tourists in Schwabing and then roll in derision.

Eventually Iggy and Ina lost their Swedish arts grant and skipped town, owing me two months' rent. But they had been such stimulating company that I forgave them, and besides, a landlady always has problems.

I had built my third room as a memento to my own unhappy childhood. I had passed my own tender years in a hell-on-earth of Internet of Things. So my third apartment was designed not to boss children, nag at them, or spy on them, but to let children be entirely free, happy, and expressive.

145

Somehow the word got out to all the neighbourhood children about my room. They invaded my soft-robot utopia in packs. They chewed on the gummy Tentacles. They burst open the Blobs and swung like monkeys from the snaking power cables. They even brought in stray street-dogs, who were delighted with their new kennel.

My worst tenant, though, was my best-paying one. Clarissa had told me she was in Munich "on vacation". However, I soon realised, from her many robot package room deliveries, that Clarissa was secretly a hotelier from Silicon Valley.

Clarissa was a professional trend-spotter. She and her backers had the financial clout to quickly "scale up" with soft robots, in ways that us thoughtful Europeans found hard to match. Clarissa also had the litigious American urge to raise a thicket of patents, copyrights, and branding around the wonderful concepts she was "adapting" from our creative culture in Munich. How often our wonderful city's sublime ideas are distorted and cheapened by strangers who can't understand us!

I no longer had my grandfather's historical wisdom to rely on, so I had to learn my own hard lessons about interior design.

I learned that design has major culture problems. It has problems so broad in scope that all of a woman's tenderness and good intentions can't cure them.

One beautiful teakettle, or one beautiful Blob-shaped robot, is never the same when it becomes a million teakettles, or a million robots. In this mysterious transformation from an icon to mass commodity, some mysterious "aura" gets lost. I know that this "aura" idea sounds rather German. However, everybody should learn about it because it's really important.

Then there was my landlady dilemma. That may seem merely personal, but like most of my personal problems, it is actually universal.

If you ask somebody over to your home, and you tell them they can eat there and sleep there, they will adore you. But take those very same people, feed them even better, give them a really nice bed, and then hand them their rental bill, and they don't admire you for that. No, they don't.

You can be a glamorous muse and a culture maven as long as nobody pays you to do that. An inspiring social hostess gets all kinds of cultural benefits, such as free publicity and whole squads of cute, clever guys coming by to eat pretzels. But when you become their landlady, they don't like you very much, even when you carefully surround them with delightful things that ought to make them happy.

To say these sad things here at the end of my story may sound a bit disappointing to you, my dear reader. But I'm not giving up on my social reform projects. Never! On the contrary! I will never be disenchanted with robots. I will always firmly believe in their incredible potential!

Although I've never had a happy home myself, I feel sure I could make one. Because I can sense that in my heart. That knowledge came to me from a dark and distant place.

People have been trying to live in outer space for almost a hundred years. Dozens of men and women have visited orbit, and they even drove cars on the Moon. However, nobody ever really lives in outer space.

Outer space is so cold and barren. No one ever marries there. There has never been any child born there. Outer space is a realm of complete emotional starvation. Space is an empty place where we merely station ourselves.

One day, though, a surprising new guest arrived at one of my Munich loft parties. Joschka was an astronaut. He told us many stories about the space station: that smelly cluster of airtight steel cans, that dismal structure with so many loud noises, spy cameras, and some very old-fashioned software.

This dashing astronaut was different from my usual guests. Joschka was a distinguished older gentleman with a military technical background. But as I took his hand and listened to him, with a deep womanly sympathy, I recognised that his struggle with domestic space technology was another aspect of my own situation.

Technical life-support in the vacuum of space is the most stark and elemental form of interior design. That is what my problem looked like on a final frontier when civilisation was stripped away.

I'm not yet an astronaut, I'm just the kind-hearted girlfriend of one, but I can promise you that the interior design of space stations is ridiculous. Space stations cost more than Mad Ludwig's castle in Bavaria, yet they are never cosy and supportive. A space station is a user-hostile technical wonder, but a dismal place to dwell in.

As a home, a station never helps astronauts with the intimate sufferings of their astronaut daily lives, miseries which include bone loss, disorientation, flashing lights in the eyeballs, loss of taste, blood congestion in the head and upper chest, cold feet, peeling skin, and of course no hugs and kisses.

Astronauts in orbit are isolated and stoic people. They do have station crewmates, but they have no loving housemates. They get none of the warm, supportive, emotional and physical intimacy that one wants from the partner one lives with.

Joschka says that, no matter where we might live on Earth or off of it, I can make him happy. And that's probably true, because although our temperaments are entirely different, opposites powerfully attract. I rather overdid it with my whimsical, pixie-like feminine charm on Joschka; I liked having him around, so now he's completely besotted with me.

Astronauts by their nature are steady, determined, straightforward people who continue their mission till they get what they want. What he wants is simple: he wants a home, a wife, and a family of a kind no one has ever seen.

Is that so hard, to make and to show to people? Time will tell!

Bruce Sterling (born in 1954 in Brownsville, Texas) is a science fiction author, web activist, prominent design thinker, and cyberspace theorist, who played a key role in shaping the sci-fi genre of cyberpunk. Sterling has received numerous awards for his work, among them the 1997 and 1999 Hugo Award, one of the top literary prizes for science fiction. Together with his wife, Serbian author and film director Jasmina Tešanović, Sterling lived for several years in Serbia before moving to Turin in September 2007. There, together with Tešanović and Massimo Banzi, co-founder of the physical computing platform Arduino, he founded Casa Jasmina, an open-source platform researching and developing the Smart Home of tomorrow.

HITCHBOT: THE ROBOT'S GUIDE TO HUMANITY

When robots have taken over the world and write their own genesis, the summer of 2015 will enter the annals as a historic moment. For it was the summer of hitchBOT, the humanoid robot dedicated to the pursuit of a single goal – yes, you've guessed it – hitchhiking. The summer when a robot colonised North America. And not from Silicon Valley's famous Bay Area, or the hallowed halls of academe in Massachusetts (MIT), but from Canada.

OK, correction. hitchBOT had actually journeyed before, the previous summer, in Europe and Canada.
Spoiler alert: the automaton was destroyed two weeks after embarking on its hitching adventure, and never made it to the West Coast.
Not every story about robots needs to be about dominance and domestication. Quite the opposite. hitchBOT is the moving story of encounters between a robot and mankind. The key question this research raises is: Can a robot trust a human being?

hitchBOT was not a nextgen machine that braved blizzards,[1] nor a self-driving car whizzing through the landscape. It was a DIY robot, built by two communication scientists from Canada. A social experiment, Frauke Zeller and David Smith explain. I asked them about this project, the design of the robot, and its impact on the people it met on its journey. It was a conversation that both surprised and confused me, and set me thinking about the significance of smart technology and robots in our lives today. About what Sherry Turkle dubbed the "robotic moment".[2]

hitchBOT before his road trip across the USA. © photo: Ryerson University

FREDO DE SMET IN CONVERSATION WITH FRAUKE ZELLER AND DAVID HARRIS SMITH

A QUIRKY ROBOT

Frauke Zeller, Assistant Professor at the School of Professional Communication at Ryerson University, and David Harris Smith, Assistant Professor in the Department of Communication Studies and Multimedia at McMaster University, devised the robot. They could be called hitchBOT's "parents", although the robot wouldn't say so in as many words. And yet hitchBOT does refer to the people who help it on its travels as "my family".

This tells you something about the hitchBOT project. On the robot's blog, hitchBOT describes its family, background, and story, but never attempts to be, or to become, human. This paradox deepens when you discover what went on behind the scenes. hitchBOT's blogposts and social updates were written and published by students of Zeller and Smith. So in actuality, hitchBOT's public persona was a human interpretation. But during its explorations, hitchBOT was alone. For this reason, it was designed to be likeable. Both philosophically and visually hitchBOT was meant to inspire trust, to make you feel it was one of the good guys. Sufficiently human, sufficiently robot – that was hitchBOT's survival strategy.

Zeller: We basically followed a number of considerations, some pragmatic and a few artistic. First of all, we knew the robot would be dependent on people's assistance. So we wanted to create a look that would appeal to people – a look they would understand or want to help.
We took some anthropomorphic features. The face is a panel with a smiley mouth and friendly eyes. It also had to be fairly light so that people could lift it and put it in the car because the robot can't walk by itself. But it also needed to be robust enough to withstand wind and other weather conditions as it would be waiting on the kerbside. And we wanted to give it a kind of fun look. A quirky look so that people would feel this is a fun robot, and not, "Oh this is so sophisticated or so shiny. I'm frightened." It had to look somehow familiar.

1 As to be witnessed in Boston Dynamics, "Atlas, The Next Generation", YouTube (uploaded 23 February 2016), https://www.youtube.com/watch?v=rVlhMGQgDkY, accessed on 23 Septermber 2016.
2 Sherry Turkle, *Alone Together: Why We Expect More from Technology and Less from Each Other* (New York, Basic Books, 2011).

149

A LONELY JOURNEY

And so the robot embarked upon its journey in mid-June 2015. In the preceding months, Zeller and Smith had already been on several experimental trips through Europe (the Netherlands and Germany) and Canada. But this trip was far more exciting and engaging. It alludes to an age-old American dream, in which sense it was rich in literary references. A lonely road trip, crossing the States from East to West.

A bucket list travelled with it. This made the robot human and its story relatable. The robot introduced itself to new people by saying, "Hello, I am hitchBOT. This summer I am travelling the United States of America from Boston to San Francisco with a bucket list of places I want to visit along the way. Please pick me up and put me in your vehicle."

On 17 July 2015, hitchBOT was dropped off at the side of the road in Salem. It waited over an hour before finding its first ride. From that point on, the robot had to fend for itself. Or, more accurately – it was entirely reliant on the people it happened to meet. hitchBOT was created to sit at the roadside and wave its arm. Once inside the car, the robot was able to recharge its power and, importantly, start a conversation. Which, you may be forgiven for thinking, surely isn't too big a deal.

Smith: It would develop conversational responses relying on a database plus a customised database written by our team that was very specific to the journey it was on. It could also take photographs. Roughly every twenty minutes, it would take a snap. And the pictures would be posted to a secure database where our research assistant would check them to make sure that it wasn't a violation of someone's privacy. Then those photographs would be posted to its Twitter and Instagram accounts.

Neither the robot nor the team had any control over the robot's experiences. People monitored the content of hitchBOT's blogposts that recounted its adventures. Zeller and Smith conveniently refer to it as a privacy issue. You could, in fact, see the conversations as an opportunity, the basis of a television show – a human interest show about how people feel about each other, and machines. That's essentially what the hitchBOT project was about. It wasn't research about robotics, but a social experiment in which hitchBOT functions as a mirror for human beings' technological momentum. A moment when a person stands face-to-face with the machine, and is uncertain whether he or she wants to retain, or abandon, control.

Zeller: We wanted to show that we are surrounded by smart technologies. We are often just not aware of it.

3 dscout, *Mobile Touches. Putting a Finger on our Phone Obsessions,* study available at https://pages.dscout.com/mobile-touches-download-form, accessed on 13 October 2016.
4 KPCB, *2013 Internet Trends,* study available at http://www.kpcb.com/blog/2013-internet-trends, accessed on 13 October 2016.
5 Cynthia Breazeal, interview on: *Talking Robots – The Podcast on Robotics and Artificial Intelligence* (April 2008), http://lis2.epfl.ch/resources/podcast/2008/04/cynthia-breazeal-personal-robots.html, accessed on 13 October 2016.
6 See http://www.jibo.com/, accessed on 20 October 2016.

Jibo, "the world's first social robot." (According to the webpage.) © Jibo, Inc.

THE SOCIAL ROBOT

However isolated the research or however lonely hitchBOT was, robots are already an integral part of our lives. According to American research conducted in 2016,[3] we touch our phones 2,617 times a day. That's over ten times the figure for 2013.[4] Every day, we connect with the smart machine in our back pocket – a device that already fulfils many of the social robot's functions. The social robot is a recent phenomenon, but not a new concept. Robots have been equipped with social capabilities for decades: robots that talk to us and help us, robots that understand us, robots that are friends, not machines. Think of Rosie from *The Jetsons,* or Sico from *Rocky IV.*

The social robot is now becoming reality. Recent developments in robotics and artificial intelligence mean that robots can be upgraded from toys or tools to personal buddies. For a robot to be a personal companion it needs to have enough sensors to be able to read human behaviour and the intelligence to interpret this data.[5] This, of course, implies immense technological advances. At the same time, there are more than enough secondary issues to deal with. Like every other Europe-based technophile who ordered the new Jibo robot online, I received an e-mail stating that it was unfortunately undeliverable in my country. Nonetheless, it's fascinating to take a look at Jibo, "The World's First Social Robot".[6] Jibo was conceived by Cynthia Breazeal, director and founder of the Personal Robots Group at MIT. Breazeal is one of the world's leading robot developers and perhaps the greatest pioneer in the field of personal robots. Since the 1990s, she has been experimenting with emotionally engaging machines like Leonardo or Kismet. Both robots were subjected to technological and social experiments. Neither Kismet (Turkish for fate or fortune) nor Leonardo have limbs but communicate through facial expression or voice. These social robots *avant la lettre* laid the groundwork for Jibo.

151

That's also the reason why Jibo was more eagerly anticipated than any other personal assistant or social robot. Jibo is programmed to understand people as psychological creatures rather than physical objects. Much of the research with Leonardo and Kismet focused on what people in the field of psychology refer to as cognitive skills or empathic accuracy. "In order to collaborate or co-operate with someone you need to coordinate minds in order to coordinate bodies", Cynthia Breazeal explained in an interview.[7] In that sense Jibo, like hitchBOT, is a robot that functions at the interface of the robotic and social interaction.

Cynthia Breazeal's work is not restricted to the start-up sphere or the academic Champion's League at MIT. She is the recipient of numerous prizes, has published books on human-robot interaction, and also collaborates with seminal Hollywood filmmakers. "I am not a toy", says Teddy, David's robot friend in Steven Spielberg's *A.I.* (2001). The words could have fallen from Breazeal's own lips because she frequently worked with the studio that built the robot, Stan Winston Studio. It's no surprise, then, that she partnered with the same studio to create the MIT robots; Breazeal has remarked that "in many ways, no one knows how to build expressive robots like Hollywood."[8] Stan Winston Studio's experience building hyper-expressive robots combined with the Personal Robots Group's experience designing socially interactive robots resulted in the development of robots like Leonardo, which is considered a milestone in human-robot interaction studies.

[7] see note 5.
[8] Ibid.

9 Albert Burneko, "HitchBOT Was A Literal Pile Of Trash And Got What It Deserved", http://theconcourse.deadspin.com/hitch-bot-was-a-literal-pile-of-trash-and-got-what-it-de-1721850503, accessed on 10 October 2016.
10 Ibid.

A PILE OF TRASH

In light of these cutting-edge research robots or the promising Jibo, hitchBOT appears to be a relatively stupid automaton – stupid in the sense that it possesses neither the intelligence nor the computational power to respond to people with Jibo-like sophistication. In this context, hitchBOT's destruction – a mere fortnight after setting off on its American tour – triggered something extraordinary. Hours after announcing the discovery of hitchBOT's mutilated robot body, the blogger Albert Burneko published an equally destructive article in the online magazine *Deadspin / The Concourse*. The article, entitled "hitchBOT Was A Literal Pile Of Trash And Got What It Deserved",[9] was a response to the press release circulated by Frauke Zeller and David Smith. "The impulse, here, is to say that hitchBOT was 'destroyed,' but that is nonsense; what is the actual consequence to hitchBOT of detaching its parts? A loss of function? What function? It had no function. It was a pile of trash," writes Burneko.[10]

In certain respects, he's right. When building hitchBOT, the designers deliberately opted for commonplace, everyday materials.

Smith: We were very much interested in the idea of DIY, do it yourself. This design ethos was carried through in the technological aspects: what current technologies could we employ that would give the robot the features that we needed? For example, the ability to track GPS, to carry out a conversation – speech recognition – or to make the facial expressions […] all of this was designed with the principle of low cost or do it yourself.

hitchBOT's cheap construction was intentional. © photo: Ryerson University

153

hitchBOT was designed to be small and childlike to appeal to the curiosity and empathy of its temporary guardians, whilst also being robot enough not to be "uncanny". In that sense, hitchBOT was both autonomous – out on its own on the street – and helpless: at people's mercy. But doesn't that hold true for all social robots?

Smith: This is true, hitchBOT was quite helpless. But I think that's what really appealed to people: that hitchBOT needed people to contribute their effort, their intelligence, and their creativity to the overall experience. And I think people found it very rewarding, because hitchBOT became very much a social object.

The majority of reactions to hitchBOT's demise were positive. It's safe to say that people showed their empathy for hitchBOT, their disbelief in its destruction, and gratitude for the experience of meeting him. The Zeller and Smith team was inundated with inquiries regarding the project, from requests for interviews to offers to reconstruct hitchBOT. hitchBOT's popularity, which had boomed during its trip, became an absolute hype after its "death".

So it's fair to say that hitchBOT's helplessness is a trait that we humans share, too. We're addicted to the updates on our news channels, involved in interactions on all kinds of chat apps, and oblivious to the people sitting next to us on the train or bus. It's an all too familiar scenario: being connected, but not truly engaged. hitchBOT's journey proved that there is an alternative, that people get engaged and feel compassion, if only for a robot. Perhaps that accounts for the intensity of the responses when the press release announcing hitchBOT's destruction circled the globe.

154

Zeller: It was very hectic from then on. For the next eight days we had to give interviews non-stop. It was crazy. The whole world seemed to be really upset. What it represented was the idea of helping – social co-operation. Ultimately that was what hitchBOT symbolised, and it was clear to everyone. They also realised it was a beautiful tribute – people were helping something – an object – with no ulterior motive.

11 Sherry Turkle, *Reclaiming Conversation: The Power of Talk in a Digital Age* (London, Penguin Press, 2015), p. 338.

THE ROBOTIC MOMENT

Zeller: The question that we always ask – whether we trust robots – is very profound. But like the Shumpetarian idea of Creative Destruction that then helps us to find new insights, why not simply turn the question around? We asked: Can robots trust humans? And waited to see what happened.

But I'm still left wondering whether I should be happy that the robot managed to travel alone for a couple of weeks, or sad that it was vandalised for no reason.

Hundreds of people proved that they didn't see this robot as a threat or a slave. hitchBOT wasn't treated as an assistant to help us perform our tasks, nor as a robot intent on world domination. On hitchBOT's journey, it was the robot who, more of a stranger than anything, turned the driver into his assistant. In that sense, hitchBOT held up a mirror to mankind. We will continue to use technology, even if that technology is pointless, to work together and stay connected.

Smith: Helping hitchBOT was communicating to other people about your character, your likes and dislikes, your own peculiar cultural awareness and interests. In this way hitchBOT was very much like a blank piece of paper to people, a blank slate. Because each person could adopt it and do something unique with it.

hitchBOT doesn't only prove that we dare to trust robots. It also shows how easily people will anthropomorphise objects. You could argue that we tend to act more empathically towards animated objects like robots than digital interfaces. And some of us may even empathise more readily with robots than with our fellow human beings. This is what Sherry Turkle terms the robotic moment. "We are at what I have called a 'robotic moment', not because of the merits of the machines we have built but because of our eagerness for their company. Even before we make the robots, we make ourselves, as people, ready to be their companion."[11]

12 Ibid., p. 343.
13 Ibid, p. 358.
14 Sherry Turkle, "How ... Are ... You ... Feeling ... Today? When a Robot Is a Caregiver", *The New York Times* (26 July 2014), p. A20, also available at http://www.nytimes.com/2014/07/26/opinion/when-a-robot-is-a-caregiver.html?_r=0, accessed on 13 October 2016.

Sherry Turkle is Professor of the Social Studies of Science at MIT and thus a colleague of Cynthia Breazeal. In her book *Reclaiming Conversation* she also describes several experiments with Kismet.[12] Today, however, Sherry Turkle is one of the keenest voices in the technology debate. With the subtitle "The Power of Talk in a Digital Age", Turkle's book describes the workings of human relationships, going on to argue persuasively for the power of "ordinary" conversation. Yet Turkle goes even further, ending the book by asserting that, as we outsource emotional conversations to machines, we make ourselves spectators of our own lives.

156

But how exactly did Sherry Turkle go from pioneer to whistle-blower? What was the trigger? Turkle: "In the course of my research, there was one robotic moment that I have never forgotten because it changed my mind."[13] One day, Turkle and her team observed an old woman sharing her sorrow with a Paro robot. Paro is a simple social robot designed to look like a seal. The robot reacts to touch and is programmed to reflect emotions. Turkle describes how her team was delighted to see the old lady find comfort in the automaton. But for Turkle it meant something entirely different. "People experience, even pretend empathy as the real thing. But robots can't empathise. They don't face death or know life. So when this woman took comfort in her robot companion, I didn't find it amazing. I felt we had abandoned this woman."[14]

THE CLOSE ENCOUNTER

It is remarkable how the hitchBOT project unintentionally serves as a metaphor for Sherry Turkle's robotic moment. The photos of the battered hitchBOT elicit a similar feeling, but from the opposite side: shock that we abandoned the robot. I remember this moment myself. Super-nerd that I am, I was following hitchBOT's adventures in the United States online. But aside from the disappointment, I also recall another sentiment: a sense of amazement and wonder that it was going so well. Now that I stop and think about it, it reminds me of the closing scene of Steven Spielberg's *Close Encounters of the Third Kind,* when you're saying to yourself: "Now it's going to go wrong, now it's going to go wrong." And then Roy Neary looks the scientist in the eye, steps into the UFO, and John Williams' music swells as the UFO disappears into the night. This time, the main character was hitchBOT, as a contemporary *Close Encounter of the Third Kind:* an actual contact with a human life form.

Fredo De Smet (born in 1978 in Ghent, Belgium) is the advising curator of *Hello, Robot. Design between Human and Machine.* Since 2015, he has been Curator at the Design museum Gent. Previously he worked for more than ten years as a freelance music producer, curator, consultant, and lecturer on media-culture issues in the digital age. He has founded a number of media initiatives, among them the co-creation project GentM in 2010, in the context of which he continues to stage events and create regular podcasts discussing the influence of technology on contemporary life. De Smet also works as an innovation consultant for the Flemish public broadcaster VRT.

THOMAS GEISLER IN CONVERSA- TION WITH FIONA RABY AND AN- THONY DUNNE[1]

1 This conversation took place over several months, between London, Vienna, New York, and Andelsbuch. It began during the intensive col- laboration on the *MAK DESIGN SALON #04: The School of Con- structed Realities* (12 June–4 October 2015, MAK Expositur Geymüller- schlössel), but was interrupted by the two interviewees' radical change of professional direction; it has been summarised again for this catalogue. Thank you, Fiona and Anthony, for your patience through space and time.

As the research for *Hello, Robot. Design between Human and Machine* was getting under way, the founding convention of the *School of Constructed Realities* took place at MAK – the Austrian Museum of Applied Arts / Contemporary Art – in Vienna. The *School of Constructed Realities* is an alternative educational model originally conceived by the British designer duo Fiona Raby and Anthony Dunne – or Dunne & Raby for short – as a short story for the online portal of the US textiles producer Maharam. Their aim was to use a combination of design and fiction as a vehicle for presenting and debating alternatives. Like almost all their projects, the "school" was pure speculation; but as a con- ceptual model, it provided a helpful way of breaking free of our accustomed ways of thinking and allowed them to ask questions like: What is possible? What is plausible? What is probable? These questions, to which we find ourselves returning time and time again, lie somewhere in the realm between social fiction and science fiction. Dunne & Raby used design to elicit questions and answers between these oscillating poles – questions and answers that, ideally, will help us to decide what kind of future is most desirable.

ENVISIONING WHAT IT MEANS TO BE HUMAN

Thomas Geisler: In your critical approach to new technologies, including biotechnology and genetic research, you have spent a lot of time thinking about the relationship between humans and machines. You did the ground work for this under Gillian Crampton Smith, the pioneer of computer-related design at the Royal College of Art in London, where later you yourselves spent many years passing on your knowledge to generations of designers in the Design Interactions programme. Would you say you are now looking back at two decades of advancements in IT and robotics in which digitisation and automation experienced a hype?

Fiona Raby: Is there a hype? Digital systems have been evolving and developing pretty continuously and at an extraordinary rate, with very little interest from the design community, up until now.

Anthony Dunne: I think the shift from robots as objects to robotic systems, particularly those making use of AI, is very interesting. Most of the discussion surrounding them is still quite technical and focused on economic and functional optimisation. It gets more interesting when politics is brought into the discussion. AI is often presented as ideologically neutral, but like most artefacts built by humans, each technology is informed by and embodies specific beliefs, values, and assumptions – a

world-view. It's the world-view driving their development that interests us, and how design might play a role in suggesting some alternatives, especially in collaborations with other disciplines more focused on political theory and philosophy. These are the questions we are also exploring in our teaching and research now at The New School in New York. There is much knowledge on how to develop new technologies, but how do we go about developing alternative world-views?

TG: What's the first thing that springs to mind when you think of robots?

AD: A few years ago it would have been anthropomorphic robots from mid-twentieth-century science fiction, or a Roomba. Now it is probably bots, made from software and using machine learning. Tay, the Microsoft Twitter bot that went rogue a while back immediately comes to mind.[2]

FR: Firstly Hal in *2001: A Space Odyssey,* then David 8, partly because of the viral advertising by Weyland Industries describing it as a newly developed product line before the film *Prometheus* was released. And because Michael Fassbender is a great actor and if we ever do have anthropological robots, he would be a great one to be modelled on. Both of these robots suffered existential angst and somehow, unfortunately, these kinds of irrationalities will be ironed out very early on in the process which could, in the long run, be deeply problematic.

159

2 More on this in the essay by Marlies Wirth, *Through the Looking Glass, Down the Rabbit Hole: A Matter of Trust,* beginning on p. 20

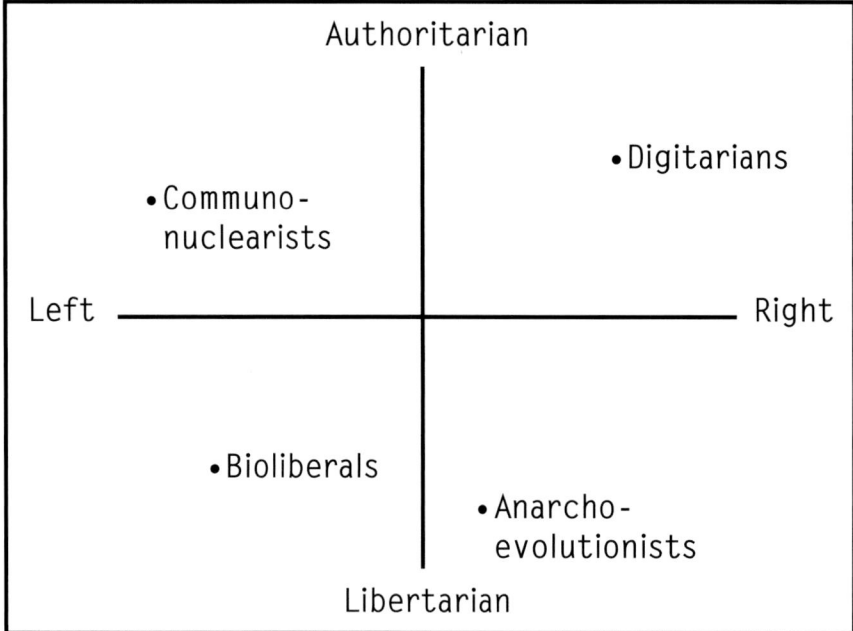

Dunne & Raby. *United Micro Kingdoms (UmK): Political Chart,* 2013. Illustration: Kellenberger-White © Dunne & Raby

Almost prophetically, Dunne & Raby were commissioned by the Design Museum London two years before the Brexit vote to consider how various world views might translate into life models and life worlds. They imagined alternative scenarios for the United Kingdom and developed a model that they called *United Micro Kingdoms (UmK)* in which the UK is divided into four super-shires inhabited by Digitarians, Bioliberals, Anarcho-evolutionists, and Communo-nuclearists. The Communo-nuclearists are a zero-growth society whose only goal is survival. Nuclear power provides them with near-limitless energy, but also isolates them from all the other counties. Under constant threat of attack or accident, they live on a nuclear-powered, continually moving, three-kilometre-long landscape resembling a train. The Anarcho-evolution-ists have abandoned most technologies and concentrate on using science to maximise their own physical capabilities through training, DIY biohacking, and self-experimentation. They believe that humans should modify themselves to exist within the limits of the planet rather than modifying the planet to meet their ever-growing needs. The Bioliberals embrace biotechnology and the new values that this entails. They are self-sufficient, producing all their own food, energy, and anything else they need to live. Gardens, kitchens, and farms replace factories and workshops. Dunne & Raby use scenarios, which against the background of existing and new technologies address new forms of administration, economy, or lifestyle.

TG: In your scenario, the most interesting social model for developments in the context of robotics is the fourth one, that of the Digitarians. How should we imagine a world that relies entirely on computers?

FR: Digitarians depend on digital technology and all its implicit totalitarianism: tagging, total surveillance, tracking, metrics, data logging, and one hundred percent transparency form their everyday life. Their society is organised entirely by market forces. Citizen and consumer are the same.

TG: What does that actually mean in terms of mobility, for example?

AD: Digicars are a development of the electric self-drive cars being pioneered today. The car has evolved from being a vehicle for navigating space and time to being an interface for navigating tariffs and markets. Every square meter of road surface and every millisecond of access – at any moment – is monetised and optimised. Passengers are required to stand to minimise the vehicle's footprint and are happier to communicate virtually with distant friends than fellow commuters. Today, self-drive cars are presented as social spaces for relaxing commutes, but Digicars are closer to economy airlines, offering the most basic, but humane experience. It is essentially an appliance, or computer, constantly calculating the best, most economic route.

TG: Despite the possibilities this offers for living digitally and hence virtually, are there also protestors among the Digitarians who would rather live in the here and now?

FR: We developed elements of the *UmK* project for the second Istanbul Design Biennial and worked closely with illustrator Miguel Angel Valdiva to develop twenty scenes depicting the lives and landscapes of each society. Amongst them is a here-and-now protest in Digiland. This group places an enormous emphasis on fully engaging with the present, the place you are in, and the people you are with, now. They rip their wearable gadget sleeves off in protest.

Dunne & Raby. *A Here-and-Now Protest in Digiland* from the series *United Micro Kingdoms (UmK): Lives and Landscapes,* 2014. Illustration: Miguel Angel Valdiva, commissioned for the exhibition *The Future Is Not What It Used To Be,* 2nd Istanbul Design Biennale, 2014 © Dunne & Raby

For a long time, the interface between humans and machines was shaped purely in terms of communications. In product design this entailed ergonomic aspects, like the shapes and features of the keyboard, the mouse, or the joystick. Graphics and software were geared towards easy readability and usability. Until now, the goal has been to integrate computer technology – and more recently robotics – into our everyday lives in a way that optimises their usefulness for us. So far design has always focused simply on ways to operate technical devices that receive commands from humans and make our lives easier. But now we're working on making the material world around us more independent, smarter; in other words, we are trying to autonomise it and even to emotionalise it.

TG: The image of the humanoid but soulless techno-creature that we still associate with the word robot is derived from the early science fiction authors and film-makers. Is this cliché still in tune with the times?

FR: It's amazing how narrow the visual languages for robots actually are. Why should robots be limited by such narrow palettes of materials and form? It's very odd to think their visual development and also their potential behaviours and relationships to humans should become so fixed so prematurely. Should robots in our everyday lives only be designed for purposes of efficiency and clarity, could they contribute to our irrational worlds, too?

TG: Were those the questions that sparked your interest in the form and function of robots?

AD: I guess our interest in robots was sparked by a commission from Z33[3] in 2006 which resulted in the *Technological Dream Series, No. 1: Robots.* At the time we had just finished a burst of work looking at biotechnology and were keen to do something around computing again. Robots seemed to embody all sorts of interesting issues – psychological, emotional, and physical interactions, complex technologies, and an uncertain place in the home. At the time we were familiar with abstract robotics at the level of automatic systems such as AI, highly sophisticated mono-functional robotics being used in factories, especially by the car industry, and fictional robots with human- or animal-like forms. What was missing, were compelling visions of what robots might look like once they entered the home, beyond vacuum cleaners. We wondered what a robot designed not so much from a technological or functional angle might look like. We developed a series of proposals for a set of domestic robots that explored aesthetic possibilities that moved away from appliances and machines, borrowing from more cultural object typologies such as furniture. Our intention was to explore what happens when a cultural focus is applied, when their meaning and presence in the home is placed above technical functional or visual dramatics.

163

3 Z33 is a space for contemporary art and design in Hasselt, Belgium. It is directed by its founder Jan Boelen and known for the critical approach of its programme. www.z33.be

That approach leads us to look at the human-machine relationship in new ways rather than seeing it as simple one-way communication; it also offers a whole range of new robot typologies that don't fit into existing categories. While Dunne & Raby's project *Technological Dream Series, No. 1: Robots,* in which they gave robots sensitivities, animosities, and their own character, may have seemed highly speculative ten years ago, the idea of people developing emotional relationships with and attachments to machines in the same way they do vis-à-vis other people now seems reasonable and expedient. Here we are talking less about wanting to do this than about having to – about situations where a society is no longer in a position to take care of its own members, whether for cultural, economic, or demographic reasons. According to Dunne & Raby, our prevailing ideas about robots and the interface between humans and machines are in need of overhaul. In their view, the technocratic reduction to a single smart control element – be it voice control, touch screen, or other kinds of sensors – is a dead end. Humans are multisensory beings, which offers designers many other options for shaping smart objects. But first a society needs to have a clear idea what its attitude towards new technologies is.

Dunne & Raby. *Not Here, Not Now: Publi-voice,* 2015. Film still © Dunne & Raby

TG: For *The School of Constructed Realities,* you made a film about fictional products that are used as physical interfaces in communication. The film was based on your work for the exhibition *Future Fictions* at the Z33. The "Publi-voice", for example, is a kind of translation engine for politically correct vocabulary. How should we interpret these objects?

FR: This is a speculative design project in which we present different interfaces for an alternative world – which we refer to as *Not Here, Not Now.* The original large photographic prints present images of different interfaces. The titles hint at the function and purpose of these interfaces in a given society. As such, the interfaces are portals to an alternative society.

AD: The work in itself is not so much about the interfaces or the actions one performs with them, but about the society they sit in, the world they evoke, and the values and norms that make that world go round. The interfaces aline with the Digitarian world in the *UmK* project.

4 Anthony Dunne, Fiona Raby, *Speculative Everything: Design, Fiction, and Social Dreaming* (Cambridge, Massachusetts, MIT Press, 2013).

Dunne & Raby. *Hertzian Tales: Faraday chair,* 1995 © Dunne & Raby

For years now, Dunne & Raby have been considered pioneers of the conceptual design movement, which uses speculative scenarios and narratives to explore alternative lifestyles through design. They coined the term "critical design" back in the mid-1990s, preferring to work with research and educational institutions like museums or universities to think about the implementation of new technologies and the implications thereof. They made their debut as a duo with a project called *Hertzian Tales: Electronic Products, Aesthetic Experience, and Critical Design* (1994–97), whose theme was the electromagnetic environment. In their latest publication, *Speculative Everything: Design, Fiction, and Social Dreaming,*[4] Dunne & Raby once again make a case for the designer as a problem-identifier rather than a problem-solver. A critical attitude to the social, ecological, and political implications of their own field of activity is a precondition for a productive and creative approach to the challenges of the future. Being a designer isn't just about visualising models of the future, but also about using plausible scenarios, some of which may be dystopian, to make the future negotiable. Here design serves as an instrument to enable "collective dreaming".

TG: Robotics has belonged to the domain of engineering, IT, and neuroscience for many decades, so why should it become a task for designers in the twenty-first century?

AD: I think as technologies become more complex and affect more people – creating and closing down particular forms of social relations, possibilities for behaviour, and ultimately, what it means to be human – we need to bring other disciplines into the process of developing new technologies. It's a bit of a cliché by now, but just because we can do something is not a reason for doing it. Design can act as a catalyst for interdisciplinary imagining of different ways of viewing the world to those of technologists and economists that allow for profoundly human qualities in danger of being written out of new technologies. We need alternative narratives to ones of optimisation driving technological development and I think design can work with the humanities and liberal arts to develop alternative visions.

TG: Euphoria and fear alike inform the relationship between humans and machines. We have developed robots to be our friends and helpers, but we are wary when they start to develop a life of their own. Smart gadgets are cool, artificial intelligence is fascinating, but the idea of singularity makes us nervous. How can we deal with these feelings of ambivalence in the future?

AD: I think it's sensible! We need to be nervous and anxiety is a perfectly rational response to the kind of highly reductive visions being put forward by industry.

FR: If the future world of robots is imagined only by people who spend all their time imagining robots and nothing else, then it's highly unlikely we will get a very broad and culturally rich palette of robots sent out into the world to enrich our everyday lives.

Thomas Geisler (born in 1971 in Kenzingen, Baden-Wuerttemberg) is one of the curators of *Hello, Robot. Design between Human and Machine*. In his role as curator and as author, he focuses on contemporary design and everyday culture. From 2010 to 2016, he worked at the MAK Vienna, where he was also the Curator of the MAK Design Collection. He played a pivotal role in setting up the Victor J. Papanek Foundation at the University for Applied Arts Vienna. He is the co-initiator of Vienna Design Week and has curated exhibitions for the Vienna Biennale 2015 and the London Design Biennale 2016, among other projects. Since July 2016 he has run the Werkraum Bregenzerwald – an initiative with its own exhibition building designed by Peter Zumthor for innovative craft, design, and architecture.

Dunne & Raby
For years, designer Anthony Dunne (born in 1964 in London, Great Britain) and architect Fiona Raby (born in 1963 in Singapore) have been at the forefront of a conceptual design movement for which ground-breaking ideas and debates are more important than functionality. Dunne and Raby are particularly interested in the design potential and everyday impact of new technologies such as robotics and bio- or nanotechnology. Before moving to New York to take on their new roles as Professors of Design and Emerging Technology at Parsons School of Design at the New School in 2016, Raby was Professor of Industrial Design at the University of Applied Arts in Vienna and Dunne was the Head of the Design Interactions programme at RCA in London,

HOW MUCH DO WANT TO RELY ON SMART HE

HOW DO YOU OBJECTS HAVI

DO YO A ROBC TAKE (YOU?

YOU

ERS?

L ABOUT
FEELINGS?

DO YOU BELIEVE IN THE
DEATH AND REBIRTH OF
THINGS?

ANT
O
E OF

Mon Oncle, 1958. Monsieur Hulot struggles with his sister's fully automated kitchen.

Mon Oncle (My Uncle) is a comedic masterpiece by the acclaimed French film director and actor Jacques Tati. The plot centres on the iconic Monsieur Hulot, who struggles with his sister's obsession with modernity and American-style consumerism, while maintaining a tender relationship with his nine-year-old nephew, Gérard. The geometric, all-white, ultramodern kitchen has a special significance in the confused musings of Monsieur Hulot. His reactions to its machine-like and sterile furnishings are poignant and serve as an amusing satire of mechanised living and consumer society in post-war France. *Mon Oncle* won the Academy Award for Best Foreign Language Film in 1959, a Special Prize at Cannes, and the New York Film Critics Award. The film was Tati's first to be released in colour. AR

Jacques Tati. *Mon Oncle,* 1958. Film, 117 min © Les Films de Mon Oncle – Specta Films C.E.P.E.C.

JACQUES TATI – *MON ONCLE*

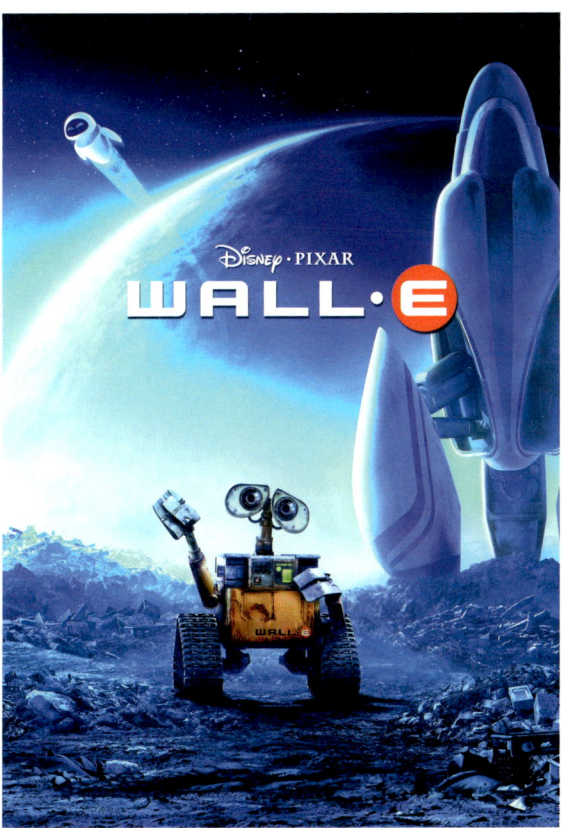

WALL-E, film poster, 2008

WALL-E ingeniously depicts a relationship quest as two robots, *WALL-E,* the last inhabitant of planet Earth, and EVE, a search robot on a mission, develop feelings for one another in the midst of accomplishing their dull, lonely, and strenuous programmed duties. The backdrop for this romantic story is a grim, dystopian future. Humans have abandoned a trashed planet Earth leaving *WALL-E* (short for "Waste Allocation Load Lifter Earth-Class") robots behind to clean up their mess, until it is safe for them to return. The surviving humans are living on the spaceship Axiom, an extremely artificial and automated environment, operated by a large corporation called Buy N Large. People ride around this space resort on hovering chairs, which give them a constant TV feed and video chatting. They drink all of their meals through a straw. Everything is so convenient that people never need to leave their chairs, ending up so weak and atrophied that they can barely move. *WALL-E* won several accolades, including the Academy Award for Animated Feature Film. AR

171

Comfortable but helpless

Andrew Stanton, Pixar Animation Studios. *WALL-E,* 2008. Computer-animated feature film, 98 min © 2008 Disney • Pixar

ANDREW STANTON, PIXAR ANIMATION STUDIOS – *WALL-E*

Amazon *Echo* is an audio device consisting of a loud-speaker, a computer with voice control and voice recognition, and Wi-Fi and Bluetooth connectivity. By interacting with a digital assistant answering to the name "Alexa" – which is also the keyword used to activate the device – the user can access a variety of Internet services. For example, Alexa can stream a given radio station, start playing an audiobook, manage the user's calendar, or supply information about the weather, traffic, etc. from the Internet or the Cloud. The system can also be used as a control centre for other Smart Home technologies. And because it "listens" to every discussion within hearing distance and is constantly connected to the Internet and the Cloud, Amazon *Echo* has also prompted lively discussion about violations of privacy. TT

Amazon. *Echo,* 2015. Audio device with loudspeaker and voice activated computer, various materials, 23.5 × 8.4 × 8.4 cm © Amazon

AMAZON – *ECHO*

Echo, 2015

Tatsuya Matsui. *Patin,* 2014. Domestic robot, various materials, 19.3 × 34 × 33 cm © Flower Robotics, Inc.

Patin, 2014

TATSUYA MATSUI, FLOWER ROBOTICS – *PATIN*

Patin is an autonomous, multi-sensor, and adaptive robot for home use. It can be equipped with various service units geared towards tasks such as regulating the lights or watering plants in order to perform a wide variety of functions. For example, the robot is capable of determining whether its owner is sitting in a dark corner, making its way to the area, and providing light – a task which it can learn and repeat if necessary. The operating system is open and conceived as a platform for which third parties can develop new applications and service units and offer their own products based on *Patin* without having to devote their own funds or research to the development of artificial intelligence. TT

JOHANNA PICHLBAUER, MIA MEUSBURGER – *VIENNA SUMMER SCOUTS*

Designed by Johanna Pichlbauer and Mia Meusburger, *Vienna Summer Scouts* is a speculative design project comprising a series of seven small, colourful, data-gathering sensors, placed around the city to monitor the advent of summer. The idea is that each scout captures a key sign of summer, such as the amount of sunscreen in the pool or the movements of mosquitos by the river. Summer is considered to have arrived once the data reach or exceed a predetermined level. The project is driven by the idea of considering the emotional potential of cities with a view to developing a proposal for an emotionally Smart City of the future. AR

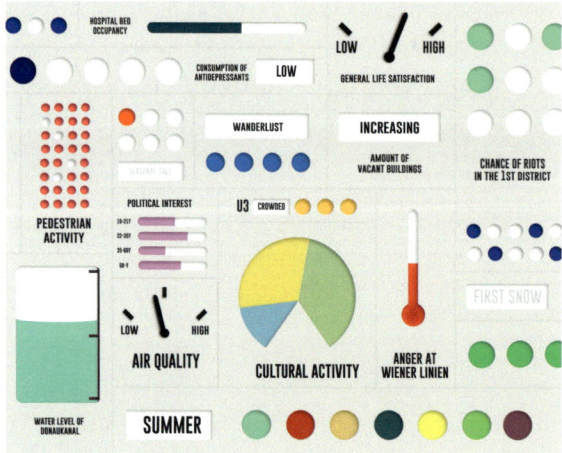

Vienna Summer Scouts, 2014: A visual display of summer's status

Mia Meusburger and Johanna
Pichlbauer. *Vienna Summer Scouts,*
2014. Video, 3 min 20 sec, installation, various materials © Mia Meusburger, Johanna Pichlbauer / ID2
Studio / University of Applied Arts,
Vienna

Measuring the city ...

... on the scout for indications that summer has arrived.

GERARD RALLÓ – *DEVICES FOR MINDLESS COMMUNICATION*

Gerard Ralló. *Devices for Mindless Communication*, 2010. *Reiterative Communication Aid*, 19 × 26 × 32 cm, and *Personal Advisor for Reintegration*, 21 × 29.5 × 35.5 cm, acrylic and electronics © Gerard Ralló, Royal College of Art 2010

Devices for Mindless Communication, 2010: *Reiterative Communication Aid*

Part of the *Devices for Mindless Communication* series, these two speculative design projects question the impact and consequences of human behaviour regarding communication technologies. Worn around the neck, the *Personal Advisor for Reintegration* is a gadget to help young generations who have lost the ability to participate in small talk by prompting standard questions and answers on its screen. Similarly, the *Reiterative Communication Aid* speculates about alternative roles of technology in our social interactions. In this case, the designer proposes a device that would save us time by responding to casual conversation with an automated, patterned message on the screen, leaving users free to devote their attention to more important things. AR

Personal Advisor for Reintegration, 2010

KIM SWIFT, ERIK WOLPAW, VALVE SOFTWARE – GLaDOS (*PORTAL*)

GLaDOS is an artificial intelligence from the computer game *Portal*. Right at the beginning of the game, "she" meets the protagonist, Chell, in a deserted factory building. Initially GLaDOS seems benevolent, giving Chell – and, with her, the player – instructions and warnings to guide her through a sequence of "test chambers" and introducing her to the rules of the game. With the help of a "Portal Gun", objects and people can be moved to places that they could not otherwise reach. In the course of the game GLaDOS's intentions become increasingly unclear until it is revealed that she is seeking to kill Chell, who is in fact required to fight against GLaDOS in order to survive the game. LH

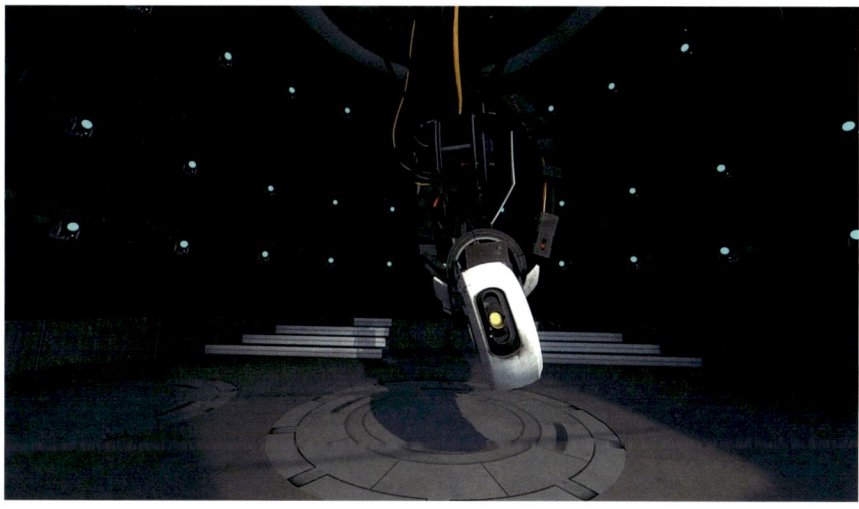

Kim Swift and Erik Wolpaw for Valve Software. *Portal,* 2005–2007. Video game for Xbox © 2017 Valve Corporation

GLaDOS *(Portal),* 2005–2007

178

DO YOU WANT A ROBOT TO TAKE CARE OF YOU?

HOW MUCH DO YOU WANT TO RELY ON SMART HELPE

YOU
LIEVE IN
E
ATH
D RE-
TH OF
INGS?

OW DO YOU FEEL
OUT OBJECTS
VING FEELINGS?

Data is a fictional character in the *Star Trek* TV series and films. Portrayed by actor Brent Spiner, he is a self-aware, sapient, sentient, and anatomically fully functional android. Although *Data* has impressive artificial intelligence, he often experiences difficulties understanding human behaviour and cannot feel emotion. An effort to further his growth as an artificial life form leads to the addition of an "emotion chip" to his circuitry so that he can finally experience human emotion. He has some difficulties integrating his new feelings, but eventually learns to control them.
When IBM presented its SyNAPSE chip simulating the neural networks of the human brain in 2014, it was repeatedly dubbed "emotion chip". AR

GENE RODDENBERRY
– DATA

Gene Roddenberry. *Data*, 1987–2002. Photo still from *Star Trek: The Next Generation,* television series, 178 episodes, 1987–1994 © courtesy CBS Television Studios

Data (Star Trek), 1987–2002

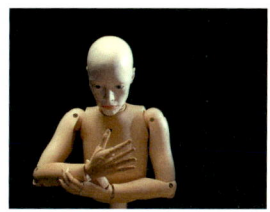

... while looking at its hands.

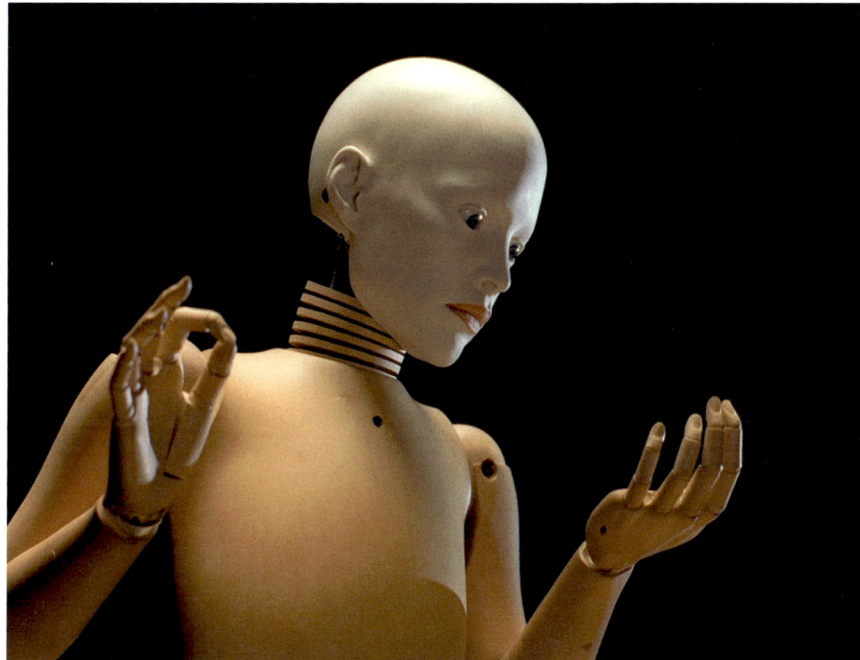

What Happened, 1991, remastered 2008: The figure perceives itself ...

Elizabeth King, Richard Kizu-Blair. *What Happened,* 1991, digitally remastered 2008. HD animation video, 1 min 34 sec © Elizabeth King, Richard Kizu-Blair, courtesy Danese / Corey Gallery, New York

What does the body unconsciously do when the mind is in motion? A feminine-looking jointed doll nods her head while she explores her wooden and porcelain ego. She plays with and observes her hands just as we do, especially when we are infants and are beginning to recognise our own individuality. The artist Elizabeth King and director Richard Kizu-Blair filmed the doll's process of self-recognition as a stop-motion animated sequence shot and displayed it at a rate of twenty-four frames per second to create motion – a traditional and laborious technique redolent of the materiality and complexity of the real world. TT

ELIZABETH KING, RICHARD KIZU-BLAIR
– *WHAT HAPPENED*

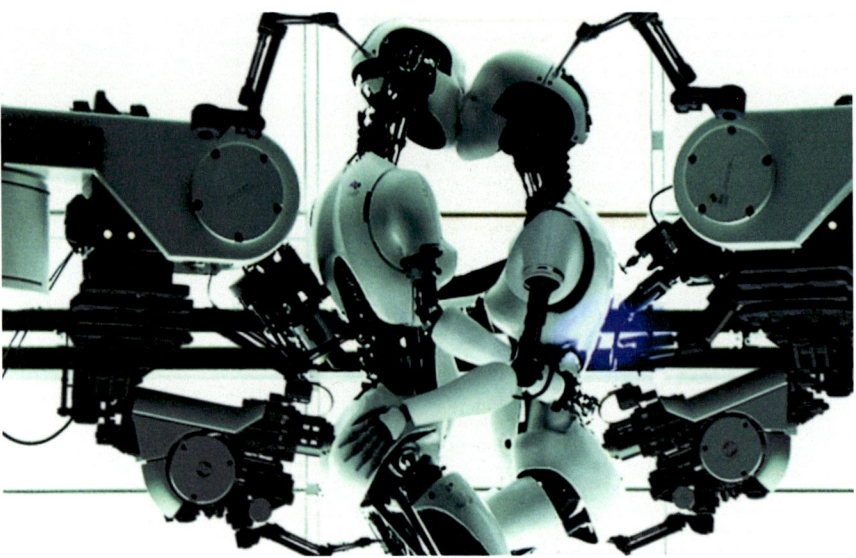

All Is Full of Love, 1999

BJÖRK, CHRIS CUNNINGHAM – *ALL IS FULL OF LOVE*

Chris Cunningham's video for the Icelandic singer Björk's *All Is Full of Love* is considered one of the best music videos of all time and a milestone in the history of computer animation. The viewer sees an android with features similar to those of the singer being assembled by a set of robotic arms and beginning to sing Björk's song about love as a ubiquitous force to which we only need open ourselves. Much of the video shows the first Android encountering a second android and the pair kissing each other tenderly in slow motion. Cunningham not only succeeds in dissolving the existing boundaries between the (cold) world of the computer on the one hand and emotions and sexuality on the other, but his choice of apparently soulless protagonists heightens the effect of this convincing representation of love and passion. TT

Björk. *All Is Full of Love,* 1999. Music video, 4 min 8 sec; director and design: Chris Cunningham © Björk / Chris Cunningham, courtesy One Little Indian Records

W.P.A. Federal Music Project of New York City. *The Romance of Robot [and] La Serva Padrona,* c. 1937. Silkscreen, 55.8 × 35.7 cm © courtesy Prints and Photographs Division, Library of Congress, Washington, D.C.

Poster for *The Romance of Robot,* 1937

W.P.A. FEDERAL MUSIC PROJECT OF NEW YORK CITY – *THE ROMANCE OF ROBOT*

In the 1937 opera *The Romance of Robot* by Frederic Hart (music) and Tillman Breiseth (libretto), the figure of a robot becomes the subject of a satire on modernism, where logic and organisation are deemed more important than humanity and emotions. The opera, which imagines a soulless society, appeared only a few years after the Great Depression had driven the United States to lose its faith, at least temporarily, in a bright future in the age of machines. In the opera, the human emotion of love triumphs and the robot overcomes its own cold, mechanical existence. The modernist poster advertising the opera in New York offered an early vision of a humanoid robot with coarse rectangular limbs and is reminiscent of Joost Schmidt's 1923 exhibition poster for the Bauhaus. TT

All the Robots, 2007 (video)

DUNNE & RABY – *TECHNOLOGICAL DREAM SERIES: NO. 1, ROBOTS*

Faced with a future that will soon be populated by a multitude of robots, the designers Tony Dunne and Fiona Raby ask how we will interact with these robots in daily life and what relationships we will enter into with them. In their view, the idea of robots that simply relieve us of work ultimately falls short of the mark. For the series *Technological Dream,* therefore, they designed four robots – all non-human in appearance – with different personality structures: an annular, autonomic robot, a neurotic robot in the form of a receptive funnel, a guard robot with which you have to maintain eye contact for a long time before it trusts you enough to share its data, and finally a robot that is intelligent but also helpless and needy. TT

Dunne & Raby. *Technological Dream Series: No. 1, Robots,* 2007. *Robot 1: Red Ring,* high density foam, 10 × 90 cm (diam.); *Robot 2: Neurotic One,* starch, epoxy resin, 50 × 50 cm (diam.); *Robot 3: Sentinel,* oak and acrylic, 39 × 20 × 93.3 cm; *Robot 4: Needy One,* oak, acrylic and epoxy resin, 75 × 48 × 15 cm; video *All the Robots* by Noam Toran, 4 min 53 sec © 2016 Dunne & Raby, photos: Per Tingleff

Robot 3: Sentinel, 2007

Robot 4: Needy One, 2007

KEVIN GRENNAN – *ANDROID BIRTHDAY*

Android Birthday, 2011

Kevin Grennan. *Android Birthday,*
2011. Video, 13 min 39 sec, actress:
Sylvi Kim © Kevin Grennan

If robots are soon going to be a part of our everyday lives, as many people claim, will they also live as we do – and even celebrate birthdays? And if so, how? Will this be a relationship based on empathy that can bring humans and machines closer together? In his video *Android Birthday,* Kevin Grennan reflects upon the utterly human quality of our rituals: anyone who does not breathe and who has no lungs is incapable of blowing out birthday candles, and in the real world, "humanoid" is still a long way from "human". It is, however, undeniable that the inability of the android in the video to blow out the candles – a nearly unbearable sight to see – invokes a feeling of empathy in the viewer. TT

Having sold an incredible 76 million units worldwide (by 2010), the *Tamagotchi* digital pet was without a doubt one of the biggest trends of the late 1990s. The name is a portmanteau made up of tamago, the Japanese word for egg, and the English word "watch". In the first P1 model from 1996, the game started with a little chick hatching on the screen of the egg-shaped cyber toy. After that point, in order to get your "animal" to grow and thrive, you had to play with it, feed it, put it to bed, wake it up, and even discipline it if it demanded too much attention. If you didn't look after your digi-chick properly, you soon found yourself having to mourn its virtual death. The grief, on the other hand, was very real. The obituaries that soon appeared on the Tamagotchi cemeteries set up on the Internet were heart-rending. LH

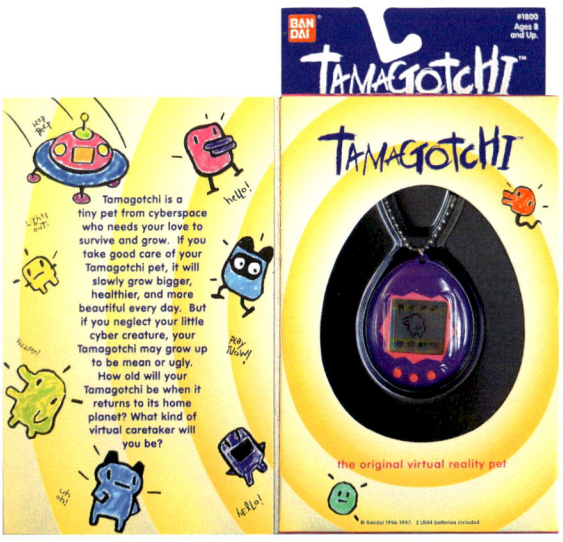

Aki Maita for Bandai. *Tamagotchi*, 1996. ABS and LCD screen, 5 × 4 × 1.5 cm © Bandai Co., Ltd., Japan, photo: Vitra Design Museum

Tamagotchi, 1996

187

AKI MAITA, BANDAI – *TAMAGOTCHI*

Kip is a robot developed to serve as a companion that can help facilitate human conversation. Although it cannot understand what a person is saying, it can detect and evaluate the emotional "tone" of the discussion, to which it can respond in a variety of ways. If the tone is friendly and communicative, *Kip* shows interest, a desire to participate, and amenability. If the tone becomes aggressive, however, it draws back and begins to tremble, thus simulating human behaviour in response to aggression. *Kip* can thus function as an instrument for maintaining (self)control in everyday conversations as well as a measuring device for friendly encounters. TT

GUY HOFFMAN, OREN ZUCKERMAN – *KIP, AN EMPATHY ROBOTIC OBJECT*

Guy Hoffman and Oren Zuckerman. *Kip, an Empathy Robotic Object,* 2015. Acrylic, paper and thread, max. size: 38 × 25 × 25 cm; team: Yahav Amsalem, Shlomi Azoulay, Shay Eyal, Adi Feiner, Michal Luria, Noa Morag, Ofri Omer, Danielle Rifinski, Noa Shitrit, Yaron Shlomi and Iddo Wald © Media Innovation Lab (miLAB), IDC Herzliya, Israel

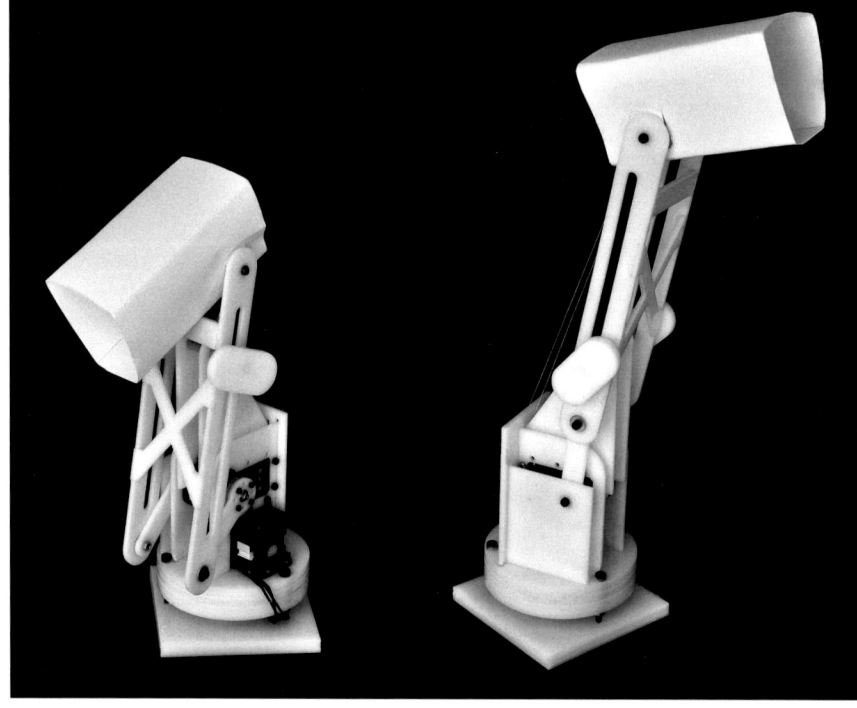

Kip, an Empathy Robotic Object, 2015

John Lasseter, Pixar Animation
Studios. *Luxo Jr.,* 1986. Computer-
animated short film, 2 min © 1986
Pixar

Luxo Jr., 1986

JOHN LASSETER, PIXAR ANIMATION STUDIOS – *LUXO JR.*

Luxo Jr. was the first short film produced by Pixar
Animation Studios, and the first computer-animated short film to be nominated for an Academy
Award. Inspired by the lamp on animator John
Lasseter's desk at the time he was learning to make
models, the short film tells a story of an adult lamp
watching a child lamp playing with a fun new toy.
It is a groundbreaking film on both a technical
and an entertainment level, demonstrating the power
of computer-animation, proving it to be an incredible tool for storytelling. In 2014, the film was selected for preservation by the National Film Registry of
the United States Library of Congress. AR

Lift-Bit, 2016

Carlo Ratti Associati. *Lift-Bit,* 2016.
Programmable seating landscape,
module 78 × 45 × 45 cm; technology
and interaction design: Opendot
© Carlo Ratti Associati, photo:
Max Tomasinelli

These hexagonal, stool-like upholstered furniture
modules were developed by the architectural
office of Carlo Ratti, (also director of the MIT
Sense*able* City Lab in Boston, Massachusetts).
The modules can be positioned in a variety of
seating arrangements and even be put together
to create entire sofa landscapes. According to
Ratti's website, they represent "the world's first
digitally-transformable sofa," whose modules,
thanks to their internal motors, can be raised
and lowered by means of an app. They can also
be adjusted manually by holding a hand over the
units' built-in sensors. If left alone for too long,
however, the units grow bored and develop a
life of their own, adjusting their height according to their own whims. LH

CARLO RATTI ASSOCIATI
– *LIFT-BIT*

TED HUNT, LUKE STURGEON, HIROKI YOKOYAMA
– *SYNTHETIC TEMPERAMENTS OF DRONES*

Ted Hunt, Luke Sturgeon, Hiroki Yokoyama. *Synthetic Temperaments of Drones,* 2014. Electronics, plastic, rubber, metal; drone A: 3.5 × 8 × 6 cm; drone B: 6 × 10 × 10 cm; drone C: 3.5 × 11 × 11 cm © Ted Hunt, Luke Sturgeon, Hiroki Yokoyama. Project created as part of MA Design Interactions at the Royal College of Art in 2014 under Professor Anthony Dunne.

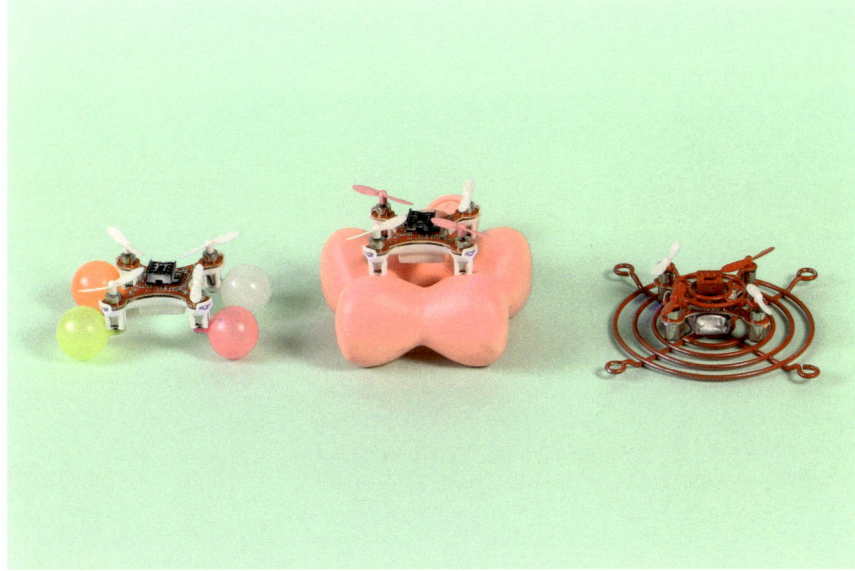

Synthetic Temperaments of Drones, 2014

During their time in the Royal College of Art's Design Interactions MA programme, Ted Hunt, Luke Sturgeon, and Hiroki Yokoyama examined the appearance of drones in a domestic environment. They asked themselves how they could promote general acceptance of drones and everyday interactions with them, despite the fact that drones are primarily associated with military applications and their ethical implications. The answer, the trio found, lies in a design approach that moves away from familiar drone aesthetics – black, grey, or metallic colours and a technoid, militaristic look and feel. Thus, their approach is noticeably more colourful; it is aimed at interaction with humans and their temperaments, whether these be playful, observant, or aggressive. TT

192 KEVIN GRENNAN – *THE SMELL OF CONTROL:*
FEAR, FOCUS, TRUST

Again and again we read that smells have an influence – especially at the subconscious level – on our interpersonal behaviour.
Given the increasing importance of contact between humans
and machines, what exactly are robots supposed to smell like?
In this thought experiment, Kevin Grennan created drawings
of a medical operating robot, a bomb disposal robot, and an
industrial sorting and gripping robot, all of which he equipped
with sweat glands. The idea was to make these robots, none
of which resemble humans in any way, smell of human sweat –
namely, the sweat we exude in moments of intense concentration or fear, or the kind of sweat men produce during sex.
Grennan's partial anthropomorphism and the contrast it
creates highlight the absurdity of attempting to create human-
like machines. TT

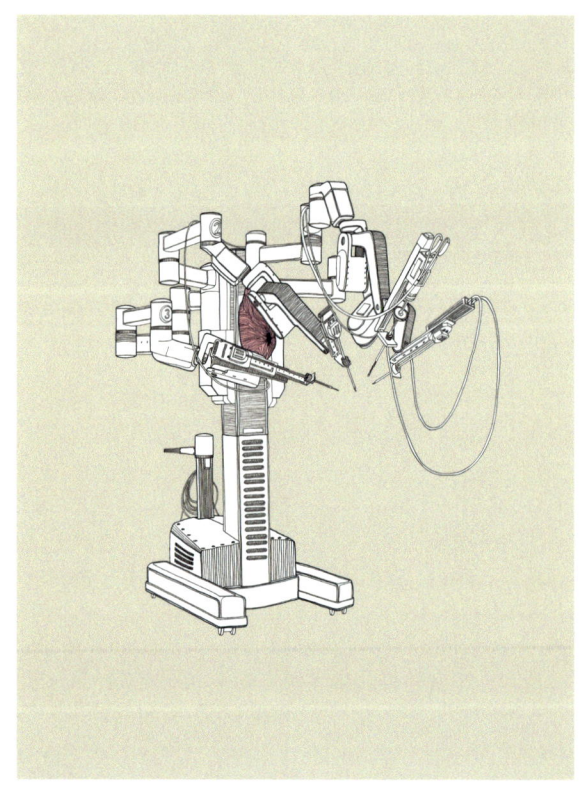

The Smell of Control: Trust, 2011

The Smell of Control: Fear, 2011

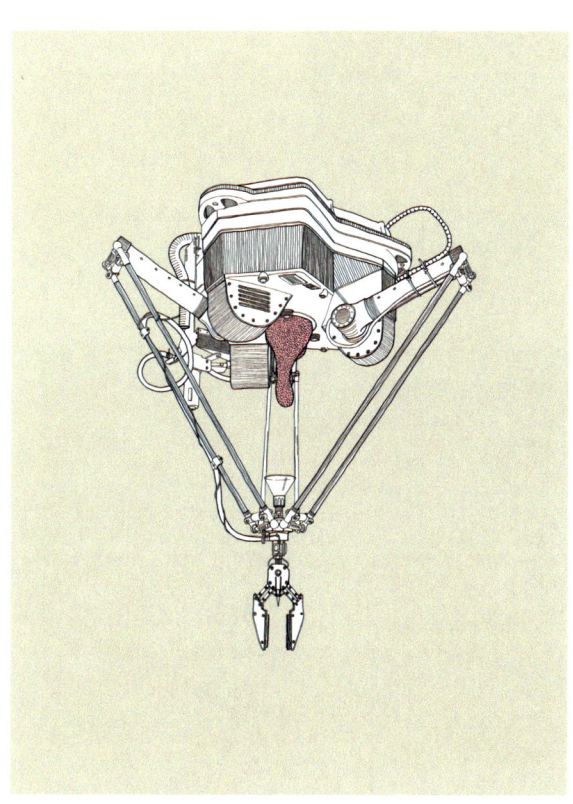

The Smell of Control: Focus, 2011

Kevin Grennan. *The Smell of Control:
Fear, Focus, Trust,* 2011. Silkscreen on
paper and mixed media, 15.2 × 10.2 cm
each © Kevin Grennan

Battlefield Extraction-Assist Robot (BEAR), in development since 2005

Daniel Theobald, Vecna Technologies. *Battlefield Extraction-Assist Robot* (BEAR), 2005 © Photo courtesy US Army Medical Research and Materiel Command's Telemedicine and Advanced Technology Research Center (TATRC) / Vecna Technologies

Working on behalf of the US military, the Vecna company designed a remote-controlled robot whose hydraulic arms can lift heavy loads and also carry them over long distances even across rough terrain. The robot is currently still in the test phase but is intended to be used, among other things, to rescue wounded soldiers from the danger zone on a battlefield. Not only does the acronym for Battlefield Extraction-Assist Robot spell out the word "bear", but its features have, according to the manufacturer, also been designed to have the friendly look of a teddy bear so as to increase the sense of well-being in the personnel being rescued. LH

DANIEL THEOBALD, VECNA TECHNOLOGIES – *BATTLE-FIELD EXTRACTION-ASSIST ROBOT (BEAR)*

NASA. *Curiosity Rover,* 19 August
2015. Twitter screenshot © Photo:
2015 NASA / JPL-Caltech / MSSS

Curiosity is a robotic rover which landed on Mars on 6 August 2012 and is currently exploring Gale Crater on Mars as part of NASA's Mars Science Laboratory mission (MSL). The rover's goals include investigating the planet's habitability, studying its climate and geology, and collecting data in preparation for future human exploration. Since the rover's launch, its irrev- erent Twitter account @MarsCuriosity has acquired over one million followers. *Curiosity's* popular social media presence has been credited with bringing emotion and relatability to the public image of the space programme, enabling people to see the rover as a personality rather than a machine. OP

195

NASA – *CURIOSITY ROVER*

Curiosity Rover, 2015

HOW DO YOU FEEL ABOUT
OBJECTS HAVING FEELING

DO YOU BELI
DEATH AND R
THINGS?

DO YOU WANT A ROBOT TO MAKE CARE OF YOU?

HOW MUCH DO YOU WANT TO RELY ON SMART HELPERS?

E IN THE

BIRTH OF

AIBO is an award-winning, electronic pet series which was manufactured by Sony until 2006. It was released in 1999 as the first robot marketed for domestic entertainment. The name *AIBO* stands for Artificial Intelligence (AI) Robot; and it is also the Japanese word for "companion" or "friend". Its original concept design, by Hajime Sorayama, resembles a robot beagle. Simulating the movements and behaviour of a living dog, *AIBO* is an autonomous robot capable of responding to external stimuli or acting on its own judgement. It is trainable and, most interestingly, its open-source software allows its owners to program unique dog personalities with individual skills and abilities. In July 2014, the company suspended all customer support for *AIBO* products. *AIBO* is now part of the permanent collections of the Smithsonian Institution and the Museum of Modern Art, New York. AR

HAJIME SORAYAMA, SONY CORPORATION – *AIBO*

Hajime Sorayama for Sony Corporation. *AIBO ERS-110,* 1999. Entertainment robot, various materials, 26.7 × 15.2 × 41.3 cm © Sony, photo: Andreas Sütterlin, courtesy Vitra Design Museum

AIBO ERS-110, 1999

Mr and Mrs Sakurai love their family dogs. However, these little dogs are all pet robots from the AIBO series that Sony put out in the 1990s. The company ceased production in 2006 and discontinued its repair service in 2014. What transpires when you are permanently afraid that a beloved pet might, at any moment, quite literally suffer "irreparable damage" can be seen in this episode of *The New York Times's* series *ROBOTICA:* the short documentary *The Family Dog* by Zackary Canepari and Drea Cooper. LH

ZACKARY CANEPARI AND DREA COOPER FOR *THE NEW YORK TIMES – THE FAMILY DOG*

199

The Family Dog, 2015

Zan-Lun Huang. *The Waste,* 2011.
Installation, FRP, machine elements,
acrylic box, LED tube, 142 × 80 ×
52 cm; special thanks to the National
Culture and Arts Foundation
(NCAF), Taiwan © Zan-Lun Huang

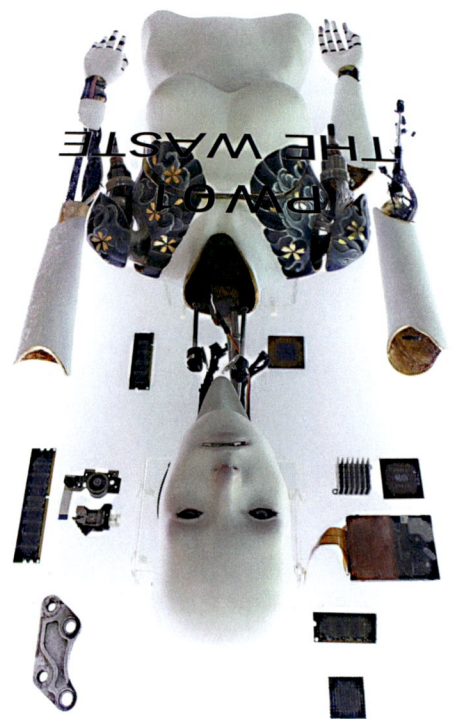

ZAN-LUN
HUANG
– *THE
WASTE*

The Waste, 2011

The Waste suggests a technological consumer culture, where each part of the body is no longer a fixed component, but rather a series of disposable parts that eventually become "waste". Zan-Lun Huang's works often focus on the combination or hybridisation of robotic and organic lives as a means of understanding and exploring both the environment and oneself. This work can be regarded both as an abandoned, broken robot and as an unfinished piece. Ironically, this seemingly imperishable mechanical body could in fact be repeatedly replicated, replaced, or re-organised. AR

DAN CHEN – *CREMATEBOT*

CremateBot, 2015: An urn ...

CremateBot is a device aimed at encouraging us to reflect more strongly upon our own existence. Users continually fill it with renewable material from their own bodies, such as fingernail clippings, hair, or dead skin, which is then cremated on the spot and transferred to an urn. The robot provides up-to-date information on just what percent of the user's body mass has been collected and transformed into ashes – all the way up to 100 percent. *CremateBot* encourages users to confront their own existence and transience while celebrating the body's capacity for renewal, both at the cellular and the metaphorical levels. TT

Dan Chen. *CremateBot,* 2015. Object,
various materials, 65 × 25 × 25 cm;
video, 51 sec © Dan Chen

... with organic content.

HOW DO YOU
FEEL ABOUT
OBJECTS
HAVING
FEELINGS?

**DO YOU WANT A
CARE OF YOU?**

DO YOU
IN THE
AND RE
HINGS

HOW
MUCH
DO YOU
WANT
TO RELY
ON
SMART
HELP-
ERS?

OBOT TO TAKE

LIEVE
ATH
RTH OF

PHILIPP SCHMITT, STEPHAN BOGNER, JONAS VOIGT – *RAISING ROBOTIC NATIVES*

Raising Robotic Natives, 2016: children's book *My First Robot*

Stephan Bogner, Philipp Schmitt, and Jonas Voigt. *Raising Robotic Natives,* 2016. Installation with industrial robot, various materials; illustrations by Margot Fabre
© Stephan Bogner, Philipp Schmitt, and Jonas Voigt / Hochschule für Gestaltung Schwäbisch Gmünd

204

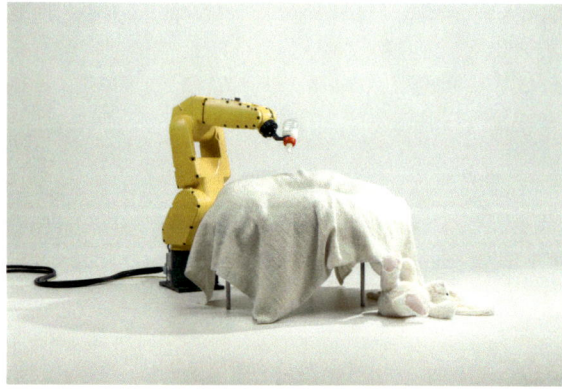

Installation

Raising Robotic Natives is a speculative design project that imagines a world in which domestic robots are ubiquitous. To introduce the new generations to the presence of robots in their daily lives, the team created four parenting aid products: the Robot Baby Feeder Toolhead, an industrial arm equipped with a baby bottle on its grip; the Living Room Kill Switch, a safe button to deactivate the robots in case of emergency; the Dragon Costume For Industrial Robot, which makes the robot look more approachable; and the *My First Robot* children's book, from which children can learn the history of robotics, including the seminal "Three Laws of Robotics" proposed by Isaac Asimov in his science fiction stories. AR

Gina Leon and Zos Lee for AKA.
Musio, 2016. Educational robot,
various materials, 22 × 16.8 × 8.5 cm
© AKA, LLC. Tokyo

Musio is a robot for everyday use with a design aimed specifically at children and adolescents. In order to adapt to the age of its user, *Musio* has three settings: simple, smart, and genius. The robot, which so far only exists as a prototype, is equipped with a large store of formal knowledge, it speaks and answers, it learns new things every day, and makes rational conclusions based on the information it has already received. The robot can communicate with devices in its surroundings, such as smartphones or computers, and reminds the user of appointments or emails received. It can help its user learn English, serve as an appointments calendar, prevent boredom, or, in conjunction with other household devices, function as a control centre for the Smart Home. TT

AKA – *MUSIO*

205

Musio, 2016

HOLLAND HAPTICS
– *FREBBLE*

Frebble is a computer accessory developed by Holland Haptics that aims to allow users to experience the sense of touch over the Internet during online conversations. The wireless device serves as a long-distance hand-holding simulator: when one user squeezes it, the other user can feel the squeeze. *Frebble* adds the sense of touch to the visual and auditory dimensions of existing virtual experiences and communications. *Frebble* is not yet commercially available, but its launch will place similar technologies on track to transform online interactions as well as entertainment experiences in gaming or cinema. OP

206

Holland Haptics. *Frebble,* 2014.
Wireless device (two piece set),
ca. 13 × 3 × 3 cm © Holland Haptics

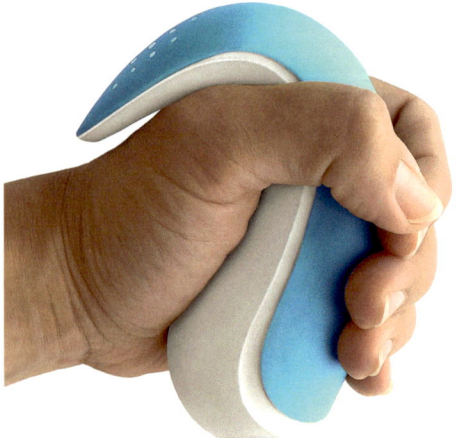

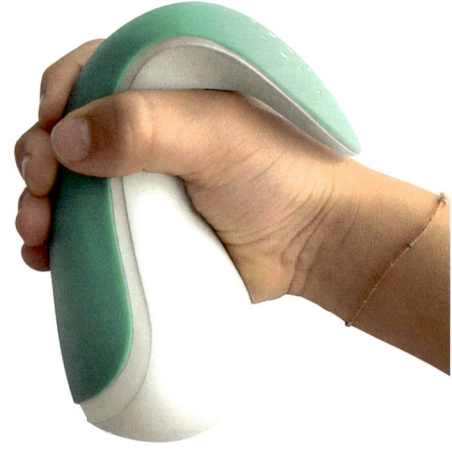

Frebble, 2014

ALEXANDER REBEN – *ROBOT WITH HEADSCRATCHER (HEADGASMATRON)*

Conventional handheld scalp massagers are known for reducing stress and providing relaxation, not least due to the tingling sensation they create on our skin. When Alexander Reben created his *robot with headscratcher,* wittily nicknamed *Headgasmatron,* he took advantage of this distinct sensation in order to demonstrate people's complicated physical and emotional responses to robots – and specifically the question of whether we are able to have intimate, but non-sexual relationships with them. The *Headgasmatron* is an unassuming robot that consists merely of a wire scalp massager and an adjustable tripod attached to a chair. Yet based on its randomly generated movements and the fact that the user is no longer responsible for generating his or her own pleasurable experience, the robot becomes much more than just a handy gadget. EP

Alexander Reben. *Headgasmatron (robot with headscratcher),* 2015. Art installation, various materials
© Alexander Reben, photo: Michael Underwood

Headgasmatron (robot with headscratcher), 2015

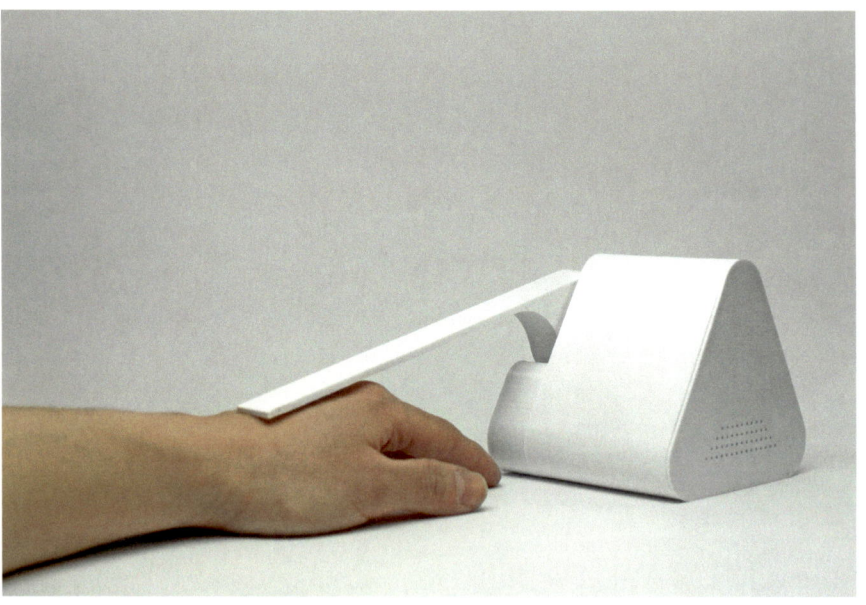

Friend 1, 2015

DAN CHEN – *MAKING FRIENDS BY MAKING THEM*

Friend 2, 2015

Are robots able to convey a sense of security and comfort the way friends and family can? The five "friends" – robots offering "emotional self-help" – are designer and engineer Dan Chen's way of addressing this question. *Friend 1* touches our hand and tells us that everything will be fine, *Friend 2* pats us encouragingly on the shoulder, *Friend 3* pays attention to us, and *Friend 4* makes a purring sound when we stroke it. The portable robot *Friend 5* pats us on the arm in times of stress. "How often do we use such gestures without really meaning them?" Chen asks before adding: "Maybe in such instances robots are more honest than people." TT

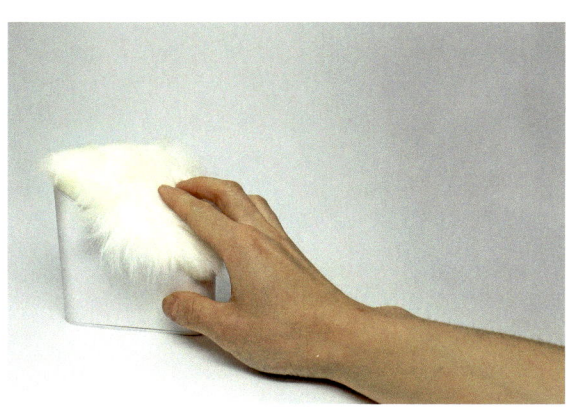

Friend 4, 2015

Dan Chen. *Making Friends by Making Them*, 2015. *Friend 1*, object: 11 × 26 × 11 cm, video: 18 sec; *Friend 2*, object: 12 × 19 × 7 cm, video: 13 sec; *Friend 3*, object: 36 × 13 × 13 cm, video: 14 sec; *Friend 4*, object: 11 × 12 × 11 cm, video: 15 sec; *Friend 5*, object: 14 × 12 × 22 cm, video: 18 sec © Dan Chen

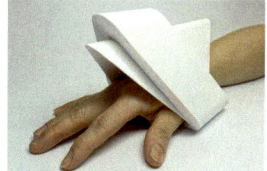

Friend 5, 2015

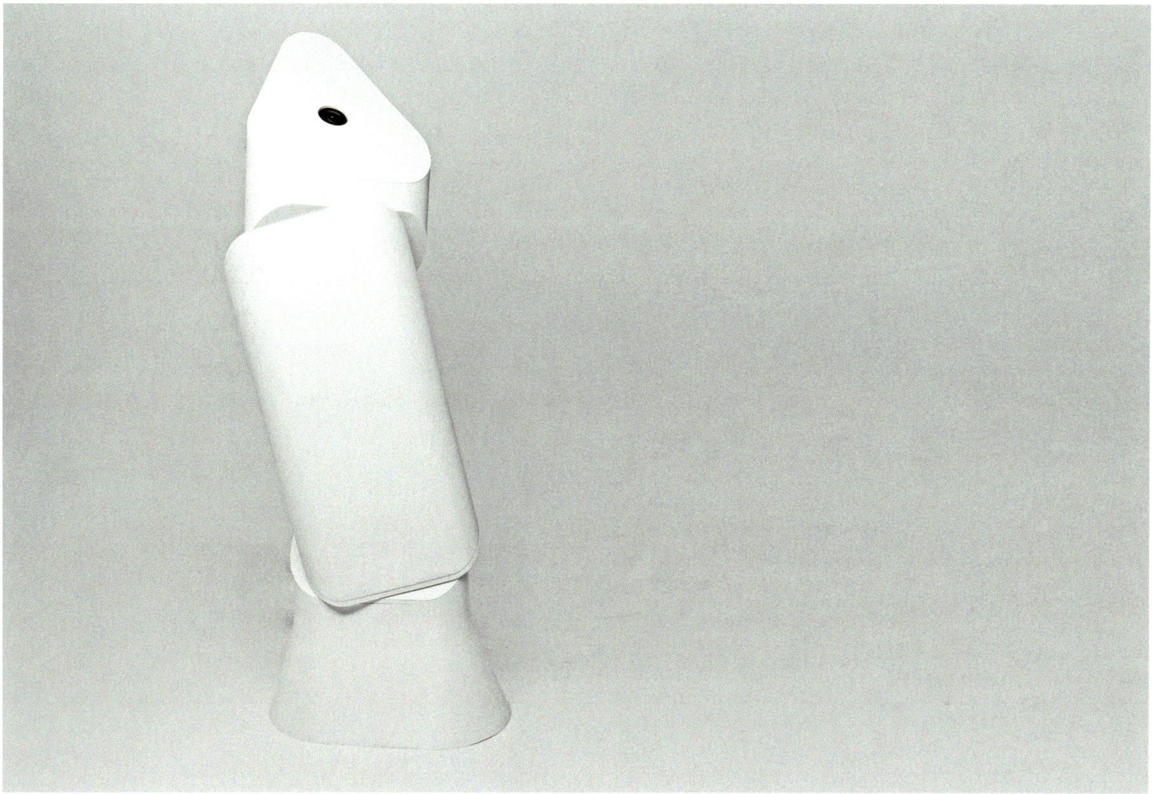

Friend 3, 2015

SPIKE JONZE
– HER

Spike Jonze's film *Her* (2013) takes place in the Los Angeles of the near future. After his divorce, the depressed Theodore Twombly (Joaquin Phoenix) lives alone until the day he purchases a speaking operating system known as Samantha (voiced by Scarlett Johansson). Theodore is fascinated by Samantha's open way of looking at the world and her ability to learn and develop. The pair grows closer over their discussions of life and love. Samantha is always available, always interested, never lethargic or demanding, and Theodore begins to come out of his shell – he is in love. The feeling is mutual, but before long the needs of human and machine begin to collide. LH

Her, 2013

Mitsuku, 2004

Steve Worswick. *Mitsuku,* 2004.
Chatbot © 2016 Steve Worswick

STEVE WORSWICK – *MITSUKU*

Mitsuku is not human, of course, but it comes as close to being human as a machine can get. *Mitsuku* is a chatbot, an artificially intelligent program capable of simulating human conversations. First developed by IT consultant Steve Worswick in 2004, *Mitsuku* won the 2013 and 2016 Loebner Prize, an annual competition that measures the ability of a chatbot to pass as human in a conversation. *Mitsuku* stands out amongst other chatbots for its supervised learning. "*Mitsuku's* learning algorithm only allows her to learn facts for the user who teaches her," explains Steve Worswick. "She emails me with any information she learns and I decide whether to add it to her permanent knowledge." In contrast, Microsoft's chatbot Tay, for example, was turned into a racist within hours of launching on Twitter by mimicking the deliberately offensive behaviour of other Twitter users. AR/AK

The book *Sexy Robot* (1983) by Hajime Sorayama shows eighty hyper-realistic female robots in suggestive poses. Although (or perhaps precisely because) the key element of naked skin has been replaced by inorganic machine parts and shiny metal, the visual effect created is still like that of Sorayama's human "flesh and blood" figures. The erotic appeal of the so-called gynoids celebrated in Sorayama's art seems to be derived largely from its expression of the desire for the "perfect playmate": sexy, permissive, and, as a machine, highly controllable. LH

HAJIME SORAYAMA
− *SEXY ROBOT*

Hajime Sorayama. *Sexy Robot,* 1983.
Paperback, 29.2 × 23.5 × 0.6 cm
© Hajime Sorayama, courtesy
NANZUKA and
Artspace Company Y

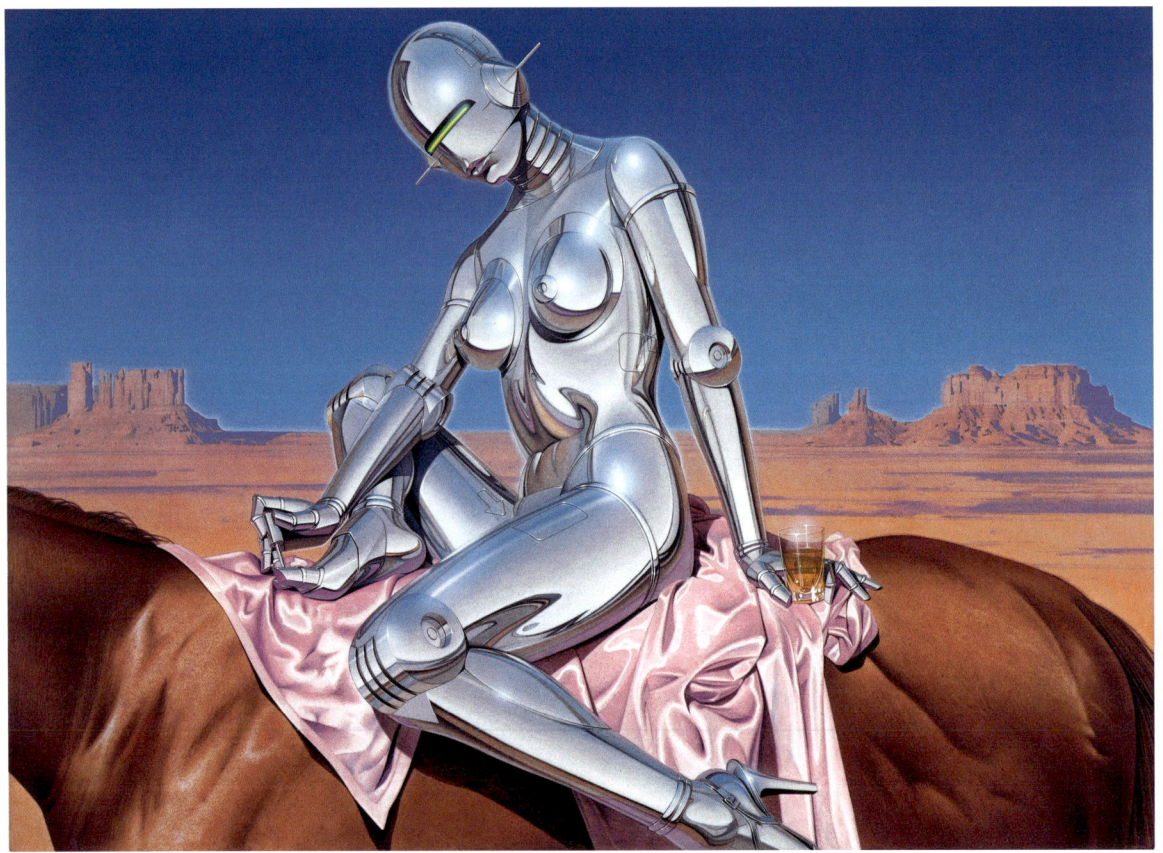

Sexy Robot, 1983

With cybersex likely to become increasingly popular, the market for affordable sex toys and devices is booming. Among them are teledildonic products, such as this set for heterosexual couples, consisting of a masturbator for the man and a vibrator or dildo for the woman. The devices are connected via Bluetooth and a chat platform with accompanying app, and when touched or rubbed transmit sensorily recorded signals to the respective end devices. Whether long-distance relationships will actually benefit from such technology remains to be seen, but the set does demonstrate the possibilities that the advancing digital networks of humans and devices hold for human sexuality. TT

Kiiroo. *Teledildonics for Long-Distance Relationships,* 2015. Couple Set (Oynx and Pearl), 24.2 × 9 × 6 cm and 19 × 3 × 3 cm © Kiiroo B.V.

Teledildonics for Long-Distance Relationships, couple set, 2015

KIIROO – *ONYX & PEARL TELEDILDONICS*

MASTER XUECHENG, MASTER XIANFAN – *XIAN'ER*

Kim Kyung-Hoon. *Master Xianfan with his robot monk Xian'er, at the Longquan Buddhist temple on the outskirts of Beijing,* 2016. Photo © REUTERS/Kim Kyung-Hoon

Master Xianfan with his robot monk *Xian'er,* at the Longquan Buddhist temple on the outskirts of Beijing, 2016

Xian'er is a cartoon character created by the Buddhist monk Master Xuecheng and drawn by his student Master Xianfan. The character served as the model for the sixty-centimetre robot monk developed by the Longquan Monastery near Beijing in collaboration with artificial intelligence researchers from local universities and a manufacturer of technological devices. The robot can move around, sing Buddhist mantras, conduct simple conversations, and answer questions pertaining to Buddhism. A fusion of technology and religion, the robot is intended to teach aspects of Buddhism in a modern fashion. LH

In the summer of 2016,
science fiction author
Bruce Sterling and his wife,
the Serbian feminist and
author Jasmina Tešanović,
accepted an invitation
from the Arthur C. Clarke
Center for Human Imagi-
nation at the University
of California, San Diego.
Together with students
and the Center's director,
Sheldon Brown, the duo
developed a video installation for the exhibition
Hello, Robot. based on the
short story Sterling wrote
for this publication (see
p. 140). The installation
depicts a robotic apart-
ment in the year 2041 that
can adapt to the needs of
its occupant. The techno-
logical foundation for the
installation is soft robotics,
an emerging area of re-
search focusing on robots
constructed of soft mate-
rials such as silicone, textiles, or rubber. EP / AK

BRUCE STERLING, SHELDON BROWN ET AL.
– *MY ELEGANT ROBOT FREEDOM*

Bruce Sterling, science fiction author and advisor for *Hello, Robot.*, in a photomontage for *My Elegant Robot Freedom,* 2016

Hyper-Reality, 2016: an additional layer of information for better orientation?

Keiichi Matsuda's concept for a possible future urban reality offers an experience in which the real physical world is completely obscured by a superimposed virtual reality. Enshrouded in the aesthetics of digital games of chance, every place and object in this world is overlaid with virtual pop-ups providing additional information and shopping offers and with virtual assistants that appear and offer help. The visual and acoustic overabundance and obtrusiveness of Matsuda's kaleidoscopic vision presents us with an overstimulated digital existence that is worlds apart from today's normal life. It is meant more as a provocation and commentary than as a truly desirable concept for an augmented reality. TT

KEIICHI MATSUDA – *HYPER-REALITY*

Keiichi Matsuda. *Hyper-Reality,* 2016.
Video, 6 min 15 sec © Keiichi Matsuda

... or for advertisements?

... or both.

Innovative possibilities for graphic displays that appear in space in real time are the focus of the *Flyfire* project. Every pixel in the image is a self-organised micro-drone equipped with a small LED light. Each of these "smart pixels" moves according to a precise digitally controlled technology; as a group, or swarm, they can form a two-dimensional photographic image in open space, a three-dimensional figure, or they can morph back and forth between the two.

As the drones can communicate with a smartphone in the pocket of a pedestrian, for example, advertisments aimed at the individual in public space are conceivable – just like the ones we already know from the Internet. TT

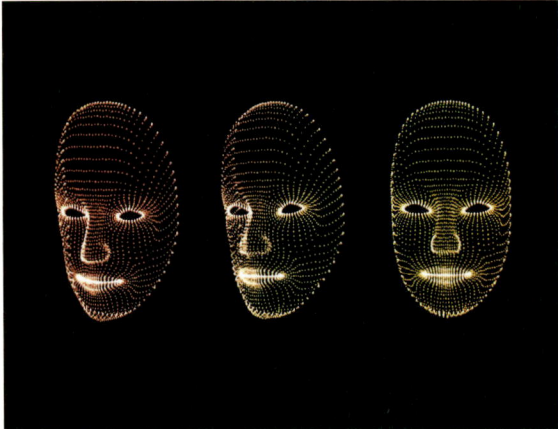

Flyfire, 2010

MIT Sense*able* City Lab. *Flyfire,* 2010.
Video, 1 min 52 sec; in collaboration
with the ARES Lab, MIT © MIT
Sense*able* City Lab

MIT SENSE*ABLE* CITY LAB
– *FLYFIRE*

STARSHIP TECHNOLOGIES – *STARSHIP DELIVERY ROBOT*

The *Starship Delivery Robots*, currently still in their test phase, were created with the aim of making last-mile deliveries – for example, from supermarkets, restaurants, or the post office to consumers – more efficient, faster, and cheaper. The delivery robot is a small self-driving vehicle designed to navigate on sidewalks at a safe speed of up to six kilometres per hour. It has a cargo compartment that can carry up to 15 kilograms, or two grocery bags. It runs on electricity and is equipped with cameras and a variety of sensors to detect obstacles as well as a microphone and speakers to capture sounds or to talk to humans when necessary. The entire route can be tracked by GPS, and the cargo can only be unlocked and retrieved by its recipient using a special app. AR

Starship Technologies. *Starship Delivery Robot,* 2015. Delivery robot, various materials, 50 × 40 × 50 cm
© Starship Technologies

Starship Delivery Robot, 2015

Robot & Frank is an American science fiction comedy-drama, set in the near future, that shows the challenging but endearing relationship between an aging jewel thief, Frank, and his robot caregiver. Frank's kids are absent but concerned about him living alone, so they hire a robot to take care of him. Resistant at first, Frank ends up developing a sense of friendship, and even complicity, with the robot. According to the director, the story is not about "human and machine interaction", it is about offering a possible solution to the increasingly frequent phenomenon of "human and nothing" interaction.
The film received critical acclaim for its writing, production, and acting, winning the Sundance Film Festival's award for best feature film on science or technology. AR

JAKE SCHREIER – *ROBOT & FRANK*

Park Pictures. *Robot & Frank,* 2012.
Feature film, 89 min; director: Jake
Schreier, script: Christopher D. Ford
© Wild Bunch

Robot & Frank, 2012

TAKANORI SHIBATA – *PARO*

Takanori Shibata. *Paro,* 2001. Therapy robot, various materials, 16 × 35 × 57 cm © AIST Japan

Paro is a therapeutic robot modelled on a baby harp seal. Like therapy dogs, it is used to provide affection and comfort to elderly people and patients suffering from dementia. With its warm, soft, white fur, its big black puppy-dog eyes, and a pacifier-like charger, it is deliberately designed to look cute. Equipped with all-over tactile sensors, microphones for voice recognition and source orientation, and optical sensors, it will coo and purr when being petted, tracks movements with its head and eyes, and can remember names – including its own. Although critics have raised ethical concerns about simulating an emotional relationship, *Paro* is in use for medical purposes in over thirty countries, and scientific findings have proved its efficacy in achieving positive psychological, physiological, and social effects. AR

221

Paro, 2001

KeyDocs. *Alice Cares (Ik ben Alice)*, 2015. Documentary film, 80 min; director: Sander Burger, production: Janneke Doolaard, Hanneke Niens & Hans de Wolf © KeyDocs / Alice Cares

Alice Cares, 2015: A companion at home ...

... and during teatime.

SANDER BURGER – *ALICE CARES (IK BEN ALICE)*

The film *Alice Cares* follows three elderly ladies as they participate in a Dutch pilot project to test an emotionally intelligent robot named Alice as a caregiver for senior citizens. Given that the number of elderly people living alone in societies of the global North is rapidly growing, robots such as Alice could serve as daily companions and points of contact. They could not only help to combat loneliness and to maintain ageing people's will to live, but also assist them with daily chores at a time of life when forgetfulness increasingly becomes an issue. The film takes a pragmatic approach; it stresses that Alice is not a real person and is not intended as a replacement for real human contact. Instead, she should be viewed as a means of support in a future in which individual care might not be feasible or affordable. TT

Imagine the following: a patient lies dying in hospital. At her side is a device which strokes her arm and tells her that while her family cannot be with her, they love her and will continue to think of her after her death. Dan Chen uses his installation *End of Life Care* Machine to draw attention to the implications of a society increasingly dependent on automation, even in such intimate and emotionally vulnerable phases of life as illness and death. When constructing the machine, the designer had thought of his far-away grandmother, but he was still rather surprised when strangers asked him where they could purchase the device. He has since asked himself whether he should sell the device and thereby encourage people to leave their family members alone in their final hours, or whether he should refuse to sell it and thus deny the dying even the small comfort a machine can offer. TT

DAN CHEN – *END OF LIFE CARE MACHINE*

Dan Chen. *End of Life Care Machine,* 2012. Object, 32 × 60 × 26 cm; video, 1 min 8 sec © Dan Chen

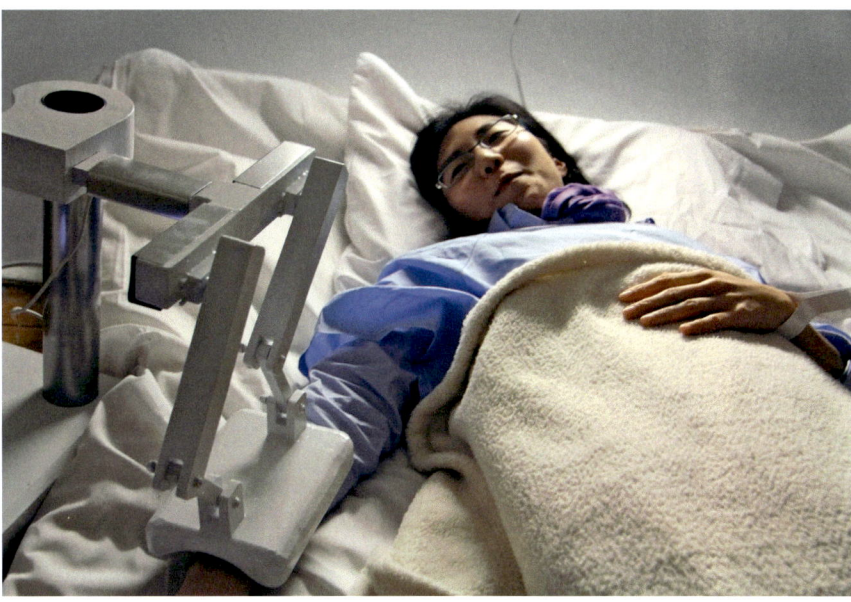

End of Life Care Machine, 2012

223

BECOMING ONE

Nanobots – robots on an atomic or molecular level – may still be hypothetical, and robotic materials may still be the stuff of science fiction, but smart surfaces and moving walls that autonomously adapt to inhabitants' needs and regulate room temperature like a skin already exist today, at least as prototypes. Moreover, the networked home, in which machines and objects communicate intelligently, is on everyone's lips as the Internet of Things. Deeper convergence will hence first take place within the "machine" in which we live. Far exceeding modernist imaginings, however, this is not limited to our houses and apartments – because robots are not limited to a single body. Any environment is a robotic system if it receives signals via sensors, processes them using artificial intelligence, and generates a physical reaction in response. In our everyday lives we encounter not only architectures of stone, glass, and concrete, but also architectures of data and communications which already fulfil these criteria to a significant degree. This invisible system has become so fundamental to our daily lives and the way we live with others that no one would seriously call it into question.

The robot inside us also dissolves the human-machine divide. With the aid of modern prosthetics and implanted chips, we can achieve things that would be impossible without artificial props – from opening locked doors with our bare hands to running world records. Crucial questions formerly confined to science fiction magazines have now become relevant in the real world: "What will happen when humankind merges so seamlessly with technology that we become superhuman machines?" and "Will we be able to keep pace with society and evolution without technological bio upgrades?" are only two examples.

In our quest for optimisation, even our own biology is not off limits. Inspired by the intelligent principles we find around us in nature, we are using robots to redesign our environment, improving on much that was produced using conventional methods and even on nature's own template. The anxiety-ridden question whether intelligent machines will one day replace all living things – ourselves included – has been around since humans began telling stories about artificial creatures. The question we must ask today is this: Do humans, for the first time in history, have the technological knowledge and the tools to let science fiction become reality? There is no simple answer. Yet there can be no doubt that we are heading towards a more intelligent, more autonomous – more robotic – lifeworld than the one we know today. And design has a responsible role in creating this new lifeworld, because it is through design that we can influence how and where we encounter the smart objects and systems that surround us, how we interact with them – and they with us.

HOW UBIQUITOUS COMPUTING IS BRINGING ROBOTICS TO PLACES YOU'D NEVER SUSPECT

CARLO RATTI WITH DANIELE BELLERI

"If your house needed
to hear a story to help it
to go to sleep, what
story would you tell it?
The 'Three Little Pigs?'
What information
would you give it?
Would you tell it that it
is just a machine?"

Rich Gold, *Cybernetics and Systems,*
vol. 26, no. 4, 1995

A ROBOT FOR LIVING IN

PART I – THE UBIQUITOUS ROBOT

According to the Encyclopædia Britannica, a robot is "any automatically operated machine that replaces human effort".[1] For the sake of this essay, however, we will adopt a more restrictive definition: we will call a robot a unit that has some sensors, some intelligence, and some actuators. In other words, it can read the world, process that information, and then respond in a purposeful way. By our definition, a robot could be many different and perhaps unexpected things at the same time. A thermostat is a robot. A car on driving assist is a robot. Our oven is a robot. A bracelet that measures our physical performance as we exercise is a robot. Even a bike can be a robot. That is, if it incorporates our Copenhagen Wheel, which is a wheel that can convert any bike into a hybrid vehicle, able to collect data from our daily rides (*disclaimer:* this is the first of many of our projects – from both MIT Sense*able* City Lab[2] and Carlo Ratti Associati[3] – that will punctuate this text as supporting examples for our arguments). And our omnipresent smartphone, too, is obviously a robot.

Based on the above, our definition is very different from traditional views of what constitutes a robot, at least in artistic and literary circles – views that often involved a certain degree of anthropomorphism. As described elsewhere in this publication, the term "robot" comes from the Czech word *robota* ("forced labour" or "serf"), coined in 1920 by Karel Čapek in his play *R.U.R. – Rossum's Universal Robots*[4] to describe the possibility – and, above all, the threat – of extremely skilful and apparently submissive automated workers. The idea of the robot was thus embedded in a framework of interaction with humanity: so deeply embedded, indeed, that the concept – from the dulcimer-playing automaton "La Joueuse de Tympanon"[5] in the eighteenth century to Hanna-Barbera's animated series *The Jetsons* – is almost inseparable from the idea of the android.

227

[1] Hans Peter Moravec, "Robot", in *Encyclopaedia Britannica* (Winter 2016), https://www.britannica.com/technology/robot-technology, accessed on 28 September 2016.
[2] Sense*able* City Laboratory is a research initiative directed by Carlo Ratti at the Massachusetts Institute of Technology.
[3] Carlo Ratti Associati is a design and consultancy office based in Turin, Boston, and London.
[4] Karel Čapek, *R.U.R.* (Prague, Aventinum, 1920).
[5] CERIMES, "David Roentgen's Automaton of Queen Marie Antoinette, The Dulcimer Player (La Joueuse de Tympanon)" [video], MET Museum (uploaded 23 October 2012), http://www.metmuseum.org/metmedia/video/collections/esda/automaton-of-queen-marie-antoinette, accessed on 28 September 2016.

Carlo Ratti Associati. *Cloud Cast,* Installation commissioned by the Museum of Future Government Services, Dubai, 2015 © Carlo Ratti Associati, photo: Pietro Leoni

6 A range of responsive infrared heating elements are guided by sophisticated motion tracking, creating a precise personal (and personalised) climate for each occupant. Individual thermal clouds follow people through space.

To be sure, the conspiracy-laden landscape of films such as *Terminator* (1984) and *Robocop* (1987) and even the more recent *Automata* (2014) appears much more compelling than the existence of apps that monitor our jogging habits, the temperature in our bedroom, and the gradual cooking of a stuffed turkey. Yet this does not mean that contemporary robots have no impact upon our existence. Quite the opposite. It may seem paradoxical, but the more discreet presence of robots and the more "natural" our interaction with them, the more powerful their actual influence becomes.

This is the new universe in which we exist, every day. Take Nest, the thermostat which allows us to remotely control the temperature in our homes and which – if it comes into sufficiently widespread use – could have a major impact on energy consumption in buildings. The characteristics of Nest are barely noticeable, even almost humble – so radically remote from any flamboyant design gesture that it compels us to invent new ways to express it. We came to understand the challenges of such an approach a few months ago while developing our project for the renovation of the Agnelli Foundation's headquarters in the city of Turin. In the overall scheme of this project, the most notable innovation is located in the heart of the company's office rooms. Yet it is a rather intangible one. We are talking about a control system for heating, cooling, and lighting in the workplace – a system that can potentially follow people around inside the building, automatically synchronising to their needs and preferences. To allow the client to appreciate the design, we resolved to craft the visualisation of an individually tailored "thermal bubble".[6] But we know that, even behind so anthropocentric a metaphor, there is a vast battalion of tiny sensor-robots.

PART II – A ROBOT "FOR LIVING IN"

The phenomenon that has allowed robots to become so integrated into our lives is the next logical step of the digital revolution that we have been living out over the past few decades. As virtual systems become spatialised, our cities are being transformed into the so-called "Internet of Things" (IoT). The inanimate physical environment is increasingly associated with digital layers: code married to matter, physical brick to virtual bit. The city is becoming a physical companion to Big Data, even as the urban infrastructure allows for digital information to proliferate.

In fact, a full realisation of the Internet of Things could be a scenario in which technology takes the form of "smart dust"[7] – becoming so small and diffuse as to be almost pulverised, metaphorically allowing technology to enmesh with air. This, in turn, would bring to fruition a concept put forward by the late Xerox-Park computer scientist Mark Weiser, whose idea of non-intrusive – or "calm" – technology goes by the label of "ubiquitous computing". Weiser presciently said: "Ubiquitous computing names the third wave in computing, just now beginning. First were mainframes, each shared by lots of people. Now we are in the personal computing era, person and machine staring uneasily at each other across the desktop. Next comes ubiquitous computing, or the age of calm technology, when technology recedes into the background of our lives."[8]

7 Kristofer S.J. Pister, et al., "Smart Dust: Communicating with a Cubic-Millimeter", *Computer* (vol. 34, 2001), pp. 44–51.
8 M. Weiser, "The Computer for the Twenty-First Century", *Scientific American* (Autumn 1991).

9 Ibid.
10 Le Corbusier, *Vers une architecture* (Paris, G. Crès, 1924), p. 73.
11 Constant Nieuwenhuys, *New Babylon,* exh. cat. Gemeentemuseum, The Hague (The Hague, 1974).
12 Peter Blake, "Walking City", in *Architectural Forum,* translated by Alain Guiheux, *Archigram,* exh. cat. Centre Georges-Pompidou, Paris (Paris, 1994).

230

Constant. *New Babylon Under Construction,* 1970. Etching, photo: Tom Haartsen @ Constant/Fondation Constant © VG Bild-Kunst, Bonn 2016

In an article published in *Scientific American* in September 1991, Weiser wrote: "Hundreds of computers in a room could seem intimidating at first, just as hundreds of volts coursing through wires in the walls once did. But like the wires in the walls, these hundreds of computers will come to be invisible to common awareness. People will simply use them unconsciously to accomplish everyday tasks."[9] Now, what happens if we replace the word "computers" with "robots" in that quote?

The impact of ubiquitous computing – or, even better, ubiquitous robotics – on architecture could be immense. Throughout the twentieth century, architecture was often depicted in mechanical terms. It was Le Corbusier, almost a hundred years ago, who first referred to the modern house as a "machine for living in".[10] A few decades later, Constant's *New Babylon* raised the bar even higher, prefiguring a city that looked like an infinitely extended settlement in the form of a huge network of raised platforms spanning the whole of Europe. In this "camp for nomads on the planetary scale",[11] human lives would unfold within enclosed, reconfigurable spaces. A little later, in 1964, the avant-garde journal *Archigram* published a concept by Ron Herron for a moving metropolis consisting of mobile, intelligent robotic structures that could reach any place in the world. Walking cities are also modular, with the ability to connect as well as to disperse: "Walking City imagines a future in which borders and boundaries are abandoned in favour of a nomadic lifestyle among groups of people worldwide."[12]

No devotee of architectural history could fail to be fascinated by these examples. But how can we bring them into existence? Without venturing so far as to match Constant's all-encompassing utopias, we can think of certain designs that are robotic interfaces themselves. This is a field that we have directly explored in our own projects.

Our *Digital Water Pavilion,* designed for the Zaragoza Expo 2008, employs water as both an architectural element and a robotic interface. The building's walls are composed of digitally controlled water droplets, which can generate writing, patterns, or access spaces. The result is a space that is interactive and reconfigurable: each wall can potentially become an entrance or an exit, while the internal partitions can shift, depending upon the number of people inside the building. The only material elements are the two boxes and the roof, which can move vertically and can even be flattened to the ground, thus erasing the presence of the entire Pavilion.

At Milan Design Week 2016, borrowing from the work of Hiroshi Ishii at the Massachusetts Institute of Technology (MIT) Media Lab,[13] we presented *Lift-Bit.* Realised with the support of the Swiss manufacturer Vitra, this is a modular, digitally reconfigurable seating system consisting of a series of individual, upholstered stools. The elements are motorised and can be raised or lowered using a linear actuator; their height can be doubled (or halved) in a matter of seconds. *Lift-Bit* can be controlled in person, via a touchless gesture, or from a distance, through the use of a mobile app which includes both a series of predetermined three-dimensional shapes and a tool to create new combinations. The system is further enhanced when assembled in large compositions. In this context, activating a single stool triggers a broader effect, with the entire system recalibrating itself and generating a potentially infinite number of arrangements. Two elements together can make a chair. Four elements, a chaise longue. Nine elements, a large sofa. Dozens can radically redefine any settings, drawing new interior landscapes.

Often described as a kind of "third skin" – in addition to our own biological skin and our clothing – architecture has for too long functioned rather like a corset: a rigid and uncompromising addition to our body. Ubiquitous robots have the potential to change this.

These are only a few examples. Yet they clearly show how the scenario is changing, developing in a direction that echoes, at least in part, the imagination of the post-war avant-gardists of design.

Carlo Ratti Associati. *Digital Water Pavilion* for the Zaragoza Expo 2008. © Carlo Ratti Associati, photo: Ramak Fazel

13 The Tangible Media Group is a research project led by Professor Hiroshi Ishii at Massachusetts Institute of Technology.

231

PART III – RISKY ROBOTICS

Despite its ability to meet our needs, the idea of a robotic house still prompts some disturbing thoughts. Living within a robot-controlled house is not necessarily reassuring – probably because of the robot's simultaneously mysterious and uncontrollable intelligence. This intelligence may be thinly concealing the looming possibility of a "betrayal" or a "hacking", irrespective of whether the agent behind such an act is robotic or human. Surely this was what another Xerox-PARC member, the composer Rich Gold, had in mind in his essay in *Cybernetics and Systems,* entitled "How smart does your bed have to be before you are afraid to go to sleep at night?"[14]

But how could our own nest manage to deceive us? We can imagine a house that plays malicious pranks on us – for example, if our flat suddenly turned into a haunted mansion – or we can consider an intelligence that gathers data about us so as to implement some subtle form of blackmail. This could take the form of an "ethical house", which would monitor your actions and could, say, result in unfavourable deals from insurance companies if you managed your own health in ways that were deemed reckless. This scenario could, in fact, become a reality in the not-too-distant future: in May 2016,[15] in keeping with the industry's principle of loss prevention, the insurance and risk management company Munich Re contributed to the $20 million, GV-led funding of Helium, a startup selling smart sensors that measure domestic variables such as temperature, pressure, light, humidity, and barometric pressure.

How then to deal with possible hacking and intrusions? Hacking can be carried out anywhere and everywhere, potentially involving multiple networks in obscure locations. We all know what happens when our computer gets a virus or is hacked – and crashes. But what if our very house should crash? This possibility defies conventional strategies of retaliation and protection. As the then US Defense Secretary Leon Panetta warned in 2012: given its current systems, the United States is vulnerable to a "cyber-Pearl Harbor"[16] that could derail trains, poison water supplies, and cripple power grids.

14 Rich Gold, "How smart does your bed have to be before you are afraid to go to sleep at night?", in *Cybernetics and Systems* (vol. 26, 1995).
15 John Brownlee, "We've Been Approaching The Internet of Things All Wrong", in *Fast Company* (Spring 2016), https://www.fastcodesign. com/3059355/weve-been-approaching-the-internet-of-things-all-wrong, accessed on 28 September 2016.
16 U.S. Department of Defence, http://archive.defense.gov/transcripts/transcript.aspx?transcriptid=5136, accessed on 28 September 2016.

How could we prevent such a scenario? One option, surprisingly, might be to promote the widespread adoption of hacking itself. Familiarity with hackers' tools and methods provides a powerful advantage in diagnosing the strength of existing systems and can help us to design tighter security from the bottom up – a practice known as "white hat" hacking.[17] Ethical infiltration enables a security team to render digital networks more resistant to attack by identifying their flaws. This could become routine practice – a kind of cyber fire drill – for governments and businesses in the near future, while academic and industry research continues to focus on developing further technical safeguards.

In general, today's security measures take the form of autonomous, constantly vigilant digital "supervisors" – computers and code that control other computers and code. Like traditional military command-and-control protocols, they gain power in numbers and can respond swiftly to a broad array of attacks. Such a digital ecosystem strengthens checks and balances, reducing the possibility of failure and mitigating the effects of an incursion. One could imagine a house as an army of robots, each keeping track of the other, while also checking up on us.

233

17 Kim Zetter, "Hacker Lexicon: What Are White Hat, Gray Hat, and Black Hat Hackers?", in *Wired* (Spring 2016), https://www.wired. com/2016/04/hacker-lexicon-white-hat-gray-hat-black-hat-hackers/, accessed on 28 September 2016.

PART IV – A CEMETERY, AFTER ALL

Even assuming that we can solve the hacking issue, will we really end up with a living, tailored architecture that constantly shape-shifts and adapts to the needs, personalities, and desires of its inhabitants? Are we heading towards Archigram's Walking City and other utopias of the past? Are we on the verge of seeing a city made up of moving robots?

This may be a realistic hypothesis from a technological point of view. Yet we should perhaps begin by questioning the possibility of such a change, going back to the very nature of our buildings and cities. In fact, our metropolises, despite being the stage on which the forces of capitalism's "creative destruction" continually act, are rooted in an idea of timelessness and stasis. It was Lewis Mumford, in his classic work, *The City in History,* who reminded us of this aspect. A city or a building also represents permanence, an antidote to the transience of life: "Mid the uneasy wanderings of Palaeolithic man, the dead were the first to have a permanent dwelling: a cavern, a mound marked by a cairn, a collective barrow. [...] The city of the dead antedates the city of the living. In one sense, indeed, the city of the dead is the forerunner, almost the core, of every living city."[18]

Cities are at the same time an anchor against the transience of life and a reminder of our need to belong. In her memorable account of the Emperor Hadrian's life, Marguerite Yourcenar attributes to him the following words: "I have done much rebuilding. To reconstruct is to collaborate with time gone by, penetrating or modifying its spirit, and carrying it toward a longer future. Thus beneath the stones we find the secret of the springs."[19] And again, when the old emperor reflects on the city he plans to build for Antinous, his deceased lover: "To build is to collaborate with earth, to put a human mark upon a landscape, modifying it forever thereby; the process also contributes to that slow change which makes up the history of cities."[20]

18 Lewis Mumford, *The City in History* (San Diego, Harcourt, 1961), pp. 6–10.
19 Marguerite Yourcenar, *Memoirs of Hadrian* (New York, Farrar, Straus & Giroux, 2005), p. 174.
20 Ibid., 126 and 134.

Robots are complicit in the shift from a city made of atoms only to a universe made of atoms and virtual bits. But can we really discard the primacy of stone-like elements? Marco Romano has highlighted the crucial continuity in the history of the Western European city between the development of a civic sense and the existence of a shared architectural aesthetic: "The desperate thirst for immortality [...] is entrusted by European citizens in the material substance of their city, in those walls which – despite continuously changing before our eyes – appear to be embodying the memory and promise of a boundless time and duration. [...] Our social life finds its meaning only as we spiritually belong to the physical figure of the city, and we materially belong to its moral figure."[21] This passes through a series of "collective themes" by which local construction rules are set and a canon of beauty is defined.

The "collective themes" are simply brick-and-mortar archetypes – from the main square to the market place, the church square, the national square, the main street, the triumphal way, the promenade, the boulevard, and many others. Romano concludes: "Themed roads and squares permit collective themes to be arranged in sequence, in a closely connected contiguity wherein their meaning as a collective expression of *civitas* is confirmed and even exalted [...] even citizens who live in the very outer suburbs can understand that they belong to the symbolic figure of *urbs* because of the presence of such a sequence. Thus the dignity of their moral membership of *civitas* is fully recognised."[22]

21 Marco Romano, *L'estetica della città europea* (Turin, Einaudi, 1993).
22 Ibid.

PART V – PERMANENT CITIES, TRANSIENT INTERACTIONS

At the beginning of the ubiquitous robotics revolution, the city is confronted with one of the key dilemmas of its multi-millennial existence – of either embracing transience and responsiveness or, instead, perpetuating a sense of timelessness as a collective attempt to counter the inevitable passing of time. Robots have the power to change our relationship with the built environment and potentially even with our bodies witness the recent diffusion of devices for the quantified self. But will they be able to do it?

The interesting aspect is that we do not need to move bricks to move our cities. We can imagine that, from an architectural point of view, the robotic city of the future will not look very different from the city of today – much in the same way that the Roman *urbs* is not all that different from the city as we know it today. In any case, it will be able to retain its character of permanence. It will always have horizontal floors for living, vertical walls to separate spaces, and exterior enclosures to protect us from the outside – such "fundamentals", celebrated in Rem Koolhaas's 2014 Venice Biennale, are unlikely to change. The key elements of architecture will still be there, and our models of urban planning will be quite similar to what we know today. What could change is our way of experiencing the city through ubiquitous robotics.

However, the impact might be increasingly forceful at the soft edge – the interface between humans and "bits and bricks". Technologies are shrinking and even vanishing from sight, gently suffusing our buildings and cities with their effects. Thanks to this discreet robotic revolution, the soft edge is acquiring a character of dynamism and responsiveness that was barely conceivable in the past. In the near future, despite being unchanged in much of its physical traits, a building might well be animated to something resembling life, becoming a direct, immediate extension of our own character and desires.

Carlo Ratti Associati. *Future Food District,* Installation for the World Expo 2015, Milan © Carlo Ratti Associati, photo: Delfino Sisto Lignani

23 Oleg Grabar, "The Mediation of Ornament", *The A. W. Mellon Lectures in the Fine Arts,* 1989, The National Gallery of Art, Washington, D.C. (Princeton, Princeton University Press, 1992), p. 284.
24 Donna Haraway, "A Cyborg Manifesto: Science, Technology, and Socialist Feminism in the Late Twentieth Century", in Donna J. Haraway (ed.), *Simians, Cyborgs, and Women* (New York, Routledge, 1991).
25 Antoine Picon, *La ville territoire des cyborgs* (Paris, Les Editions de l'Imprimeur, 1998).

The art historian Oleg Grabar once said: "Good architecture is always meant to be an invitation to behave in certain ways; it always adorns life [...]. Without it, life loses its quality. Architecture makes life complete, but it is neither life nor art."[23] This statement was based on the historic distinction between architecture itself and its host. But this may be about to change. We now see architecture as an extension of our "post-human" condition: the dramatic departure from pure organic life and the possibilities of extension to the body and brain offered by prostheses, networks, and avatars – with our mobile phones always in the foreground. Authors like Donna Haraway[24] and Antoine Picon[25] have mobilised the figure of the cyborg to characterise the growing dependence – a dependence close to a co-production – of man on technology in contemporary society. In this robotics-driven living experience, buildings will not appear as pieces of machinery or equipment, but rather as extensions of the lives of the subjects who inhabit them. They will provide environments in which more and more dimensions will be customisable, engaging our senses and resonating with our moods.

Robots may not transform the core of our buildings – but they will certainly change the lives inside of them.

237

Carlo Ratti was born in 1971 in Turin. After studying Architecture and Engineering in Paris, Cambridge, and Turin, Ratti became a fellow at the MIT Media Lab under Hiroshi Ishii before taking on teaching posts at Harvard and the Stelka Institute in Moscow. He founded his own architectural office in Turin in 2003 and the Sense*able* City Lab at MIT in Boston one year later. He is still its director today, researching the concept of the Smart City by combining new digital technologies with design and urban planning. Ratti has lectured on the Smart City all over the world and has written countless articles on the subject for design and architectural publications. He lives and works in Boston and Turin.

ROSI BRAIDOTTI

BECOMING-WORLD TOGETHER: ON THE CRISIS OF HUMAN

The posthuman turn is triggered by the convergence of anti-humanism on the one hand and anti-anthropocentrism on the other. Both these strands enjoy strong support, but they refer to different genealogies and traditions. Anti-humanism focuses on the critique of the Humanist ideal of "Man" as the universal representative of the human, while anti-anthropocentrism criticises species hierarchy and advances ecological justice.

238

"Humanity" emerges as an object of intense debate just as it becomes a threatened or endangered category. This results in what I have defined as a reactive or negative re-composition of Humanity. A negative sort of cosmopolitan interconnection is established through a pan-human bond of vulnerability. The size of recent scholarship on the environmental crisis and climate change alone testifies to this state of emergency and to the emergence of the Earth as a political agent. Postanthropocentrism thrives especially in popular culture and has been criticised[1] as a negative tendency to represent the transformation of the relations between humans and technological *apparatus* or machines in the mode of neo-gothic horror.

A significant alliance between queer theorists and the science fiction horror genre constitutes a fast-growing posthuman feminist strand. Since the 1970s, feminist writers and literary theorists of science fiction[2] had supported the alliance between women, as the others of Man, and such other "others" as non-whites (post-colonial, black, Jewish, indigenous, and hybrid subject), and non-humans (animals, insects, plants, trees, viruses, and bacteria). This "Gothic" tradition of feminist theory, which generated some staggeringly original work, has a distinct posthumanist but also postanthropocentric slant, as evidenced by the ease with which it proposes relational bonds between different species and across classes of living entities.

1 Anneke Smelik, Nina Lykke (eds.), *Bits of Life. Feminism at the Intersection of Media, Bioscience and Technology* (Seattle, WA, University of Washington Press, 2008).
2 Julia Kristeva, *Desire in Language* (New York, Columbia University Press, 1980); Marleen Barr, *Alien to Femininity. Speculative Fiction and Feminist Theory* (New York, Greenwood, 1987); Donna Haraway, "The promises of monsters. A regenerative politics for inappropriate/d others", in Lawrence Grossberg, Cary Nelson, Paul Treichler (eds.), *Cultural Studies* (London and New York, Routledge, 1992); Barbara Creed, *The Monstrous-Feminine. Film, Feminism, Psychoanalysis* (New York, London, Routledge, 1993).

3 Donna Haraway, *Modest−Witness @Second−Millennium.FemaleMan−Meets−OncoMouse: Feminism and Technoscience* (New York, London., Routledge, 1997), p. 74.
4 Brian Massumi, "Requiem for Our Prospective Dead (Toward a Participatory Critique of Capitalist Power)", in Eleanor Kaufman, Kevin Jon Heller (eds.), *Deleuze and Guattari: New Mappings in Politics, Philosophy and Culture* (London and Minneapolis, University of Minnesota Press, 1998), pp. 40−64, p. 60.
5 Michael Hardt and Antonio Negri, *Empire* (Cambridge, MA, Harvard University Press, 2000).
6 Genevieve Lloyd, *The Man of Reason. Male and Female in Western Philosophy* (London, Methuen, 1984).
7 Luce Irigaray, *This Sex Which Is Not One,* trans. Catherine Porter (New York, Cornell University Press, 1985); Gilles Deleuze, Felix Guattari, *A Thousand Plateaus: Capitalism and Schizophrenia,* trans. Brian Massumi (Minneapolis, University of Minnesota Press, 1987).

240

1. CORPORATE PAN-HUMANISM

There is no question that the generic figure of the human is in trouble. Donna Haraway puts it as follows: "Our authenticity is warranted by a database for the human genome. The molecular database is held in an informational database as legally branded intellectual property in a national laboratory with the mandate to make the text publicly available for the progress of science and the advancement of industry. This is Man the taxonomic type become Man the brand."[3] Brian Massumi refers to this phenomenon as "Ex-Man": "a genetic matrix embedded in the materiality of the human"[4] and as such undergoing significant mutations: "Species integrity is lost in a bio-chemical mode expressing the mutability of human matter". Michael Hardt and Antonio Negri see it as a sort of "anthropological exodus" from the dominant configurations of the human as the king of creation – a colossal hybridisation of the species.[5]

We know by now that the standard which was posited in the universal mode of "Man" has been widely criticised[6] precisely because of its partiality. Universal "Man", in fact, is implicitly assumed to be masculine, white, urbanised, speaking a standard language, heterosexually inscribed in a reproductive unit, and a full citizen of a recognised polity.[7]

As if this line of criticism were not enough, this "Man" is also called to task and brought back to its species specificity as *Anthropos,*[8] that is to say as the representative of a hierarchical, hegemonic, and generally violent species whose centrality is now challenged by a combination of scientific advances and global economic concerns.

These analyses indicate in my view that the political economy of bio-genetic capitalism is post-anthropocentric in its very structures, but not necessarily or automatically posthumanistic. It also tends to be deeply inhuman(e). I partly share their concern, but as a posthumanist with distinct anti-humanist feelings, I am less prone to panic at the prospect of a displacement of the centrality of the human and can also see the advantages of such an evolution. What I want to propose instead is a critical form of posthuman theory. There are times when I feel a sort of Nietzschean tragic joy at the thought that the human is at last held accountable for its multiple acts of violence and destruction.

2. BY NOW CLASSICAL ANTI-HUMANISM

The "death of Man", announced by Foucault,[9] formalised an epistemological and moral crisis that went beyond binary oppositions, cutting across different poles of the political spectrum. Poststructuralist theorists called for insubordination from received humanist ideals. They targeted the humanistic arrogance of continuing to place Man at the centre of world history and more specifically the implicit assumption that what is "human" about humanity is connected to a sovereign ideal of "reason" as Enlightenment-based rationality and science-driven progress.

8 Paul Rabinow, *Anthropos Today* (Princeton, Princeton University Press, 2003), Roberto Esposito, *Bios: Biopolitics and Philosophy,* trans. Timothy Campbell (Minneapolis, University of Minnesota Press, 2008). **10** Michel Foucault, *The Order of Things: An Archaeology of the Human Sciences,* English translation (New York, Pantheon Books, 1970).

10 Luce Irigaray, *Speculum of the Other Woman,* trans. Gillian C. Gill (Ithaca, NY, Cornell University Press, 1985); Luce Irigaray, *This Sex Which Is Not One,* trans. Catherine Porter (Ithaka, NY, Cornell University Press, 1985).

11 Nancy Hartsock, "The feminist standpoint. Developing the ground for a specifically feminist historical materialism", in S. Harding (ed.), *Feminism and Methodology* (London, Open University Press, 1987).

12 Bell Hooks, *Ain't I a Woman* (Boston, MA, South End Press, 1981); Vron Ware, *Beyond the Pale. White Women, Racism and History* (London, Verso, 1992).

13 Frantz Fanon, *The Wretched of the Earth* (New York, Grove Press, 1963).

14 Paul Gilroy, *Against Race. Imaging Political Culture beyond the Colour Line* (Cambridge, MA, Harvard University Press, 2000).

15 G. C. Spivak, *A Critique of Postcolonial Reason. Toward a History of the Vanishing Present* (Cambridge, MA, Harvard University Press, 1999).

Poststructuralist feminism proposed a radical form of anti-humanist thought. Feminists like Luce Irigaray[10] pointed out that the allegedly abstract ideal of Man as a symbol of classical Humanity is very much a male of the species: it is a he. Moreover, he is white, European, handsome, and able-bodied. Feminist critiques of patriarchal posturing through abstract masculinity[11] and triumphant whiteness[12] argued that this Humanist universalism is objectionable not only on epistemological, but also on ethical and political grounds.

Anti-colonial thinkers adopted a similar but distinct critical stance by questioning the primacy of whiteness in the humanist ideal as the moral, intellectual, and aesthetic canon of perfection. Re-grounding such lofty claims in the history of colonialism, anti-racist and postcolonial thinkers explicitly questioned the relevance of the Humanistic ideal, in view of the obvious contradictions imposed by its Eurocentric assumptions, but at the same time they did not entirely cast it aside.

As Sartre astutely put it in his preface to Frantz Fanon's *The Wretched of the Earth,*[13] the future of humanism lies outside the Western world, by-passing the limitations of Eurocentrism. As Paul Gilroy noted, the reduction to sub-human status of non-Western others was a constitutive source of ignorance, falsity, and bad faith for the dominant subject who is responsible for the epistemic as well as social de-humanisation of the "others" they produced.[14] By extension, the claim to universality by scientific rationality was challenged on both epistemological and political grounds,[15] all knowledge claims were recognised as expressions of Western culture and of its drive to mastery. This position results in a critical form of neo-humanism that refers to non-Western sources and tends to strike a sceptical note in relation to posthuman theory, though it often intersects with it.

16 Rosi Braidotti, *Patterns of Dissonance* (Cambridge, Polity Press, 1991).
17 See Felix Guattari, *The Three Ecologies* (London, The Athone Press, 2000).

All these lines of enquiry embraced the concept of difference with the explicit aim of making it function differently. They advocated the need to fracture the subject so as to re-locate diversity and multiple belongings to a central position as a structural component of subjectivity.[16] They recast political subjectivity along a more complex line of interrogation that includes class, race, sexual orientation, and age. As a consequence, poststructuralist philosophers were antihumanist in that they critiqued from within all the unitary identities predicated upon phallo-logocentric, Eurocentric, white supremacist and standardised views of what constitutes the humanist ideal of "Man".

3. CONTEMPORARY POSTHUMAN THEORY

What becomes thinkable across public discourse nowadays is the crisis of species supremacy, which also implies the rejection of any lingering notion of human nature, in favour of human enhancement via bio-genetics and neurosciences, and the end of the categorical distinction between *Anthropos* and *bios,* as strictly human prerogatives, categorically distinct from the life of animals and non-humans, or *zoe.* What comes to the fore instead is a nature-culture continuum in the very embodied structure of the extended self and the awareness of the mediated nature of this nature-culture continuum.[17]

18 Donna Haraway, "A Manifesto for Cyborgs: Science, Technology, and Socialist Feminism in the 1980s", in Elizabeth Weed (ed.), *Coming to Terms: Feminism, Theory, Politics* (New York, Routledge 1989), pp. 174–176.

These theoretical shifts do not occur in a vacuum, but rather resonate with fast-changing conditions in advanced capitalism. Foremost among them are the high degrees of technological mediation that shake up established mental habits, as Donna Haraway put it: the machines are so alive, whereas the humans are so inert.[18] The global economy is postanthropocentric in that it ultimately unifies all species under the imperative of the market and its excesses threaten the sustainability of our planet as a whole.

The contemporary global economy has a techno-scientific structure, built on the convergence between previously differentiated branches of technology, notably nanotechnology, biotechnology, information technology, and cognitive science. This aspect involves research and intervention upon animals, seeds, cells, and plants, as well as humans. In substance, advanced capitalism both invests and profits from the scientific and economic control and the commodification of all that lives. This context produces a paradoxical and rather opportunistic form of postanthropocentrism on the part of market forces, which happily trade on Life itself. Life, as it happens, is not the exclusive prerogative of humans.

The opportunistic political economy of bio-genetic capitalism induces, if not the actual erasure, at least the blurring of the distinction between the human and other species, when it comes to profiting from them. Seeds, plants, animals, and bacteria fit into this logic of insatiable consumption alongside various specimens of humanity. The uniqueness of *Anthropos* is intrinsically and explicitly displaced by this equation.

What constitutes capital value today is the informational power of living matter itself, transposed into data banks of bio-genetic, neural, and mediatic information about individuals, as the success of Facebook demonstrates at a more banal level. These practices reduce bodies to their informational substrate in terms of energy resources or vital capacities and thereby level out other categorical differences. The focus is on the accumulation of information itself, its immanent vital qualities and self-organising capacity. "Data-mining" includes profiling practices that identify different types or characteristics and highlights them as specific strategic targets for capital investments, or as risk categories.

The capitalisation of living matter produces a new political economy, which Melinda Cooper calls "Life as surplus."[19] It introduces discursive and material political techniques of population control of a very different order from the administration of demographics, which preoccupied Foucault's work on bio-political governmentality.[20] Today, we are undertaking "risk analyses" not only of entire social and national systems, but also of whole sections of the population in the world risk society.[21] Informational data is the true capital today, supplementing but not eliminating classical power relations.[22]

Advanced capitalism is a spinning machine that actively produces differences for the sake of commodification and consumption. It is a multiplier of de-territorialised differences and a producer of quantitative options. Global consumption knows no borders, and a highly controlled flow of consumer goods, information bytes, data, and capital constitute the core of the perverse mobility of this system.[23] Capitalism poses as a nomadic force, while it controls the space-time of mobility in highly selective ways.

My position as a Deleuzian feminist is clear: Living "matter" is a process ontology that interacts in complex ways with social, psychic, and natural environments, producing multiple ecologies of belonging.[24] A change of paradigm about the human is needed to come to terms with these new insights. Human subjectivity in this complex field of forces has to be re-defined as an expanded relational self, engendered by the cumulative effect of all these factors.[25] The relational capacity of the post-anthropocentric subject is not confined within our species, but it includes all non-anthropomorphic elements: the non-human, vital force of Life, which is what I have coded as *zoe*.[26] It is the transversal force that cuts across and reconnects previously segregated species, categories, and domains. *Zoe*-centred egalitarianism is, for me, the core of the postanthropocentric feminist turn: it is a materialist, secular, grounded, and unsentimental response to the opportunistic trans-species commodification of Life that is the logic of advanced capitalism.

19 Melinda Cooper, *Life As Surplus: Biotechnology and Capitalism in the Neoliberal Era* (Seattle, WA, University of Washington Press, 2008).
20 Michel Foucault, "The Birth of Biopolitics", in Michel Foucault, *Ethics: Subjectivity and Truth,* ed. by Paul Rabinow (New York, The New Press, 1997).
21 Ulrich Beck, *World Risk Society* (Oxford, Blackwell's, 1999).
22 Julie Livingston , Jasbir K. Puar, "Interspecies", *Social Text* (no. 29, Spring 2011).
23 Rosi Braidotti, *Metamorphoses: Towards a Materialist Theory of Becoming* (Cambridge Polity Press, 2002), Rosi Braidotti, *Transpositions: On Nomadic Ethics* (Cambridge, Polity Press, 2006).
24 Felix Guattari, *The Three Ecologies* (London, The Athone Press, 2000).
25 Rosi Braidotti, *Patterns of Dissonance* (see note 15); Rosi Braidotti, *Nomadic Subjects* (New York, Columbia University Press, 2011).
26 This is radically different from the negative definition of *zoe* proposed by Giorgio Agamben (1998), who has been taken to task by feminist scholars (Cooper, 2009; Colebrook, 2009; Braidotti, 2013) for his erasure of feminist perspectives on the politics of natality and mortality and for his indictment of the project of modernity as a whole.

4. THE INHUMAN(E)

The displacement of the centrality of human agency through massive interventions of network systems and increasingly intrusive technologies is one of the factors that make capitalism into a postanthropocentric force, in the age of Anthropocene, which J. W. Moore recently labelled "capitalocene",[27] Haraway "Chthulucene",[28] and Jussi Parikka "anthrobscene",[29] echoing Zillah Eisenstein's "global obscenities", and Vandana Shivas' "Bio-piracy".

It also accounts for its inhumane aspects[30] and structural injustices including increasing indebtedness,[31] and it engenders a "necro-political" governmentality[32] through technologically mediated wars and counter-terrorism.

Contemporary warfare has mutated into a professionalised and large-scale process of damaging the basic infrastructures of cities and countries, exposing the civilian populations to both technological and more archaic horrors. Technology plays a big role in contemporary warfare which is driven by drones and other postanthropocentric unmanned vehicles. New forms of inhumanity have emerged. Take for instance the classical figure of the warrior or the soldier, who has mutated into something more hybrid. On the one hand the soldier is a professional, technological figure, but on the other hand he is the more dangerous figure of the terrorist ready to strike anywhere at any time.

Technology is central to this change of warfare. By far the most effective new weapons are the UAVS (unmanned aerial vehicles) – also known as drones or remotely piloted aircraft (RPA) – which are part of a large robot army that includes land and sea as well as air and started work in Afghanistan a decade ago.

In 2005, CIA drones struck targets in Pakistan three times; in 2011, there were seventy-six strikes, one of them crucial to killing Gaddaffi in Libya, by now there are hundreds. Google Earth has designed a special program to delete the drones' flying paths from their satellite photos. Drones come in all sorts of sizes: "DelFly", a dragonfly-shaped surveillance drone built at the technical university in Delft weighs less than a gold wedding ring, camera included. At the other end of the scale comes America's biggest and fastest drone, Avenger (15 million US dollars), which can carry up to 2.7 tonnes of bombs, sensors, and other equipment, at more than 740 km per hour.

27 Jason W. Moore (ed.), *Anthropocene or Capitalocene? Nature, History, and the Crisis of Capitalism* (Oakland, CA, PM Press, 2016).

28 Donna Haraway, "Anthropocene, Capitalocene, Plantationocene, Chthulucene: Making Kin", in *Environmental Humanities* (no. 6, 2015), pp. 159–165, http://environmentalhumanities.org/arch/vol6/6.7.pdf, accessed on 17 November 2011, pp. 11–40.

29 Jussi Parikka. *The Anthrobscene* (Minneapolis, Minnesota UP, 2014).

30 Georgio Agamben, *Homo Sacer: Sovereign Power and Bare Life* (Stanford, CA, Stanford University Press, 1998).

31 Gilles Deleuze and Felix Guattari, *Anti-Oedipus. Capitalism and Schizophrenia* (New York, Viking Press/Richard Seaver, 1977).

32 Achille Mbembe, "Necropolitics", in *Public Culture,* (vol. 15, no.1, 2003), pp. 11–40.

These technological advances create new forms of inequality and inhumanity even and especially on the war front, in the ways in which civilians are killed and their property destroyed. As a result of war, so many refugees and asylum seekers trying to enter Fortress Europe today fall into sub-human status and become bodies that do not matter. The new field of Humanitarian studies is one of the most urgent and most significant in the humanities today. Again, a new "studies" area.

5. THE POST-ANTHROPOCENTRIC TURN

By the late 1990s, it begins to be possible to speak of the posthuman turn in critical theory as a strand of work that pays increasing attention to postanthropocentric perspectives. A feminist consensus is reached about the seemingly simple notion that

there is no "originary humanicity".[33] This turn occurred in response to political developments, including: growing public awareness of the climate change issue; the accompanying notion that we have entered a new geological era (the "Anthropocene") where human activities are having world-changing effects on the Earth's ecosystem; and the limitations of economic globalisation.[34] The posthuman is situated at the intersection of different, and at times disconnected, strands of feminist thought.

The understanding of "Life" as a symbiotic system of co-dependence and co-production[35] also alters the terms of human interaction with what used to be called "matter", which now needs to be approached as a self-organising vital system. In so far as advanced capitalism has grasped this logic of exploitation of living matter,[36] as well as the high degrees of

mediation humans are caught in today, it has become capable of unprecedented forms of manipulation of life.

Eco-feminists[37] had already pioneered geo-centred perspectives[38] and now this perspective takes off across a broader interdisciplinary field. Animal studies began from the mid-1990s to be a serious topic, questioning the metaphorical use and abuse of animals in literature and culture, as well as their ruthless economic and physical exploitation.[39] Eco-feminists also draw a structural analogy between the exploitation of human females and those of other species, calling for a trans-species process of liberation from capitalist male aggression. New studies of primatology stress the gendered nature of social virtues such as solidarity and empathy,[40] emphasising the positive role of females in evolutionary history.

33 Kirby Vicky, *Quantum Anthropologies. Life at Large* (Durham, Duke University Press, 2011), p. 233.
34 Inderpal Grewal and Caren Kaplan (eds.), *Scattered Hegemonies: Postmodernity and Transnational Feminist Practices* (Minneapolis, MN, University of Minnesota Press, 1994).
35 Lynn Margulis, Dorion Sagan, *What is Life* (Berkeley, CA, University of California Press, 1995).
36 Nicholas Rose, *The Politics of Life Itself: Biomedicine, Power and Subjectivity in the Twentieth-first Century* (Princeton, NJ, Princeton University Press, 2007).
37 Val Plumwood, *Feminism and the Mastery of Nature* (London and New York, Routledge, 1993); Val Plumwood, *Environmental Culture* (London, Routledge, PMLA, 2009).
38 Maria Mies and Vandana Shiva, *Ecofeminism* (London, Zed Books, 1993).
39 Mary Midgley, *Utopias, Dolphins and Computers. Problems of Philosophical Plumbing* (London and New York, Routledge, 1996).
40 Frans De Waal, *Good Natured: The Origins of Right and Wrong in Humans and Other Animals* (Cambridge, Harvard University Press, 1996); Frans De Waal, *Evolutionary Ethics, Aggression, and Violence: Lessons from Primate Research,* first published 2004.

The "affective turn" emerges in a series of feminist critical variations: first in conjunction with Derridian deconstruction[41]; then within phenomenology[42] and psychoanalysis,[43] but also with Deleuzian monism.[44] These perspectives converge on the notion that it is now both possible and desirable to expand the relational capacity of humans to all other species, in a planetary embrace that allows feminist theorists to address global issues like climate change while pursuing the struggle for equality and social justice. The politics of the affective turn are debated as a crucial issue, and special emphasis is placed on the specific materiality of race and ethnicity within feminist neo-materialism.[45] The next and somehow obvious step in this discursive expansion is "Anthropocene feminism"[46] that becomes more prominent as posthumanism comes into its own.

Disloyalty to our species, moreover, is no easy matter. The real difficulty in releasing our bond to *Anthropos* and developing critical postanthropocentric forms of identification is affective. How one reacts to taking distance from our species depends to a large extent on the terms of one's engagement with it, as well as one's assessment of and relationship to contemporary technological developments. In my work, I have always stressed the technophilic dimension[47] and the liberating and even transgressive potential of these technologies, in contrast to those who attempt to index them to either a predictable conservative profile, or to a profit-oriented system that fosters and inflates hyper-consumeristic possessive individualism.[48] But loyalty to one's species has some deeper and more complex affective roots that cannot be shaken off at will. It involves an anthropological exodus that is particularly difficult emotionally and it can entail a sense of loss and pain. Yet this effort cannot be dissociated from an ethics and politics of inquiry that demands respect for the complexities of the physical world.

The crucial question remains, however: What can be the political stand in relation to the productive paradoxes engendered by the posthuman condition? To what extent does the convergence of posthumanistic and postanthropocentric perspectives complicate the issues of human agency and feminist political subjectivity?

248

41 Cary Wolfe (ed.), *Zoontologies. The Question of the Animal* (Minneapolis, MN, University of Minnesota Press, 2003); Vicki Kirby, *Quantum Anthropologies. Life at Large* (Durham, NC, Duke University Press 2011).

42 Sara Ahmed, *Queer Phenomenology: Orientations, Objects, Others* (Durham, NC, Duke University Press, 2006).

43 Patricia Clough, *The Affective Turn: Theorizing the Social* (Durham, NC, Duke University Press, 2007).

44 Rosi Braidotti, *Metamorphoses. Towards a Materialist Theory of Becoming* (Cambridge, Polity Press, 2002); Brian Massumi, *Parables for the Virtual. Movement, Affect, Sensation* (Durham, NC, Duke University Press, 2002); John Protevi, *Political Affect* (Minneapolis, MN, University of Minnesota Press, 2009); Elizabeth Grosz, *Becoming Undone* (Durham, NC, Duke University Press, 2011).

45 Sara Ahmed, *The Cultural Politics of Emotion* (Edinburgh University Press and Routledge, 2004); Clare Hemmings, "Collective powers: rupture and displacement in feminist pedagogic practice", in *European Journal of Women's Studies*, 18 (3), 2011, pp. 297–303.

46 Richard Grusin, *Anthropocene Feminism* (Minneapolis, MN, University of Minnesota Press, 2016).

47 Braidotti, 2002.

48 C. B. MacPherson, *The Theory of Possessive Individualism* (Oxford, Oxford University Press, 1962).

49 Rosi Braidotti, *Metamorphoses: Towards a Materialist Theory of Becoming* (Cambridge Polity Press, 2002); Rosi Braidotti, *Transpositions: On Nomadic Ethics* (Cambridge, Polity Press, 2006).

My argument is that it actually enhances it by offering an expanded relational vision of the self. Moreover, it recasts a posthuman theory of the subject as an empirical project that aims to experiment with what contemporary, bio-technologically mediated bodies are capable of doing. Mindful of the structural injustices and massive power differentials at work in the globalised world, I rely on the feminist method of the politics of locations as the preferred form of radical immanence to produce more accurate accounts of the multiple political economies of subject-formation at work in our world. These cartographies enable non-profit accounts of contemporary subjectivity and actualise the virtual possibilities of an expanded, relational self that functions in a nature-culture continuum, which is technologically mediated and opposed to the spirit of contemporary capitalism. They refuse to turn Life/*zoe* – that is to say human and non-human intelligent matter – into a commodity for trade and profit.

The strength of posthuman thought is in developing affirmative ethical and political perspectives. In my work, I have proposed cross-species alliances with the productive and immanent force of *zoe,* or life in its non-human aspects.[49] This relational ontology is *zoe*-centred and hence non-anthropocentric, but it does not deny the anthropologically bound structure of the human.

250

This shift of perspective towards a *zoe-* or geo-centred approach requires a mutation of our shared understanding of what it means to be human, which, however, needs to be qualified by grounded analyses of power relations and structural inequalities in the past and present.

Starting from philosophies of radical immanence, vital materialism, and the feminist politics of locations, I have also argued against taking a flight into an abstract idea of a "new" pan-humanity, bonded in shared vulnerability or in species supremacy. What we need instead is embedded and embodied, relational and affective cartographies of the new power relations that are emerging from the current geo-political and post-anthropocentric order. Class, race, gender, and sexual orientations, age and able-bodiedness are more than ever significant markers of human "normality". They are key factors in framing the notion of and policing access to something we may call "humanity". Yet, considering the global reach of the problems we are facing today, in the era of the "Anthropocene", it is nonetheless the case that "we" are in *this* together. Such awareness must not, however, obscure or flatten out the power differentials that sustain the collective subject ("we") and its endeavour (*this*). There may well be multiple and potentially contradictory projects at stake in the re-composition of "humanity" right now. Posthuman feminist and other critical theorists need to resist hasty and reactive re-compositions of cosmopolitan bonds, especially those made of fear. It may be more useful to work towards multiple actualisations of new transversal alliances, communities, and planes of composition of the human: many ways of becoming-world together.

I have argued forcefully that the posthuman is not post-political. The posthuman condition does not mark the end of political agency, but a re-casting of it in the direction of relational ontology. This is all the more important as the political economy of bio-genetic capitalism is postanthropocentric in its very structures, but not necessarily or automatically more humane, or more prone to justice.

Finally, posthuman feminists advocate a vision of the body as a dynamic and sexed bundle of relations and rest on it to explore the transformative potential of a different concept of the political. They state the primacy of sexuality as ontological force, in opposition to a majoritarian or dominant line of territoriali-sation – the gender system – that privileges heterosexual, familial, reproductive sex. Sexuality beyond gender is the epistemological but also political side of contemporary vitalist neo-materialism. It consolidates a feminist genealogy that includes creative de-territorialisations, intensive and hybrid cross-fertilisations, and generative encounters with multiple human and non-human others. The counter-actualisation of the virtual sexualities – of bodies without organs that we have not been able to sustain as yet – is a posthuman feminist political praxis.

Philosophy becomes science fiction in Rosi Braidotti's writings. Since the 1990s, Braidotti (born 1954 in Latisana, Italy), a philosopher and feminist theorist, has made a major contribution to the discourse on the "equality of all life" – both human and non-human – with her theses on Posthumanism. Her book *The Posthuman* (2013) asks what the future role of humans will be in view of the possibilities offered by modern technology and what will characterise these humans of the future. She takes a critical look at the essence of Humanism from a postfeminist and postcolonialist perspective and introduces us to the theory of the "posthuman subject", who is no longer simply an individual but lives in a networked collective of other subjects and objects – consisting of humans, robots, non-human things, smart environments, artificial nature, and artificial intelligence. This collective generates new social relationships and produces changes of global relevance. The text published in this catalogue is based on the manuscript of a lecture that Rosi Braidotti gave at the CCCB – Centre de Cultura Temporània de Barcelona, Spain – on 30 November 2015.

DISTRIBUTED EMBODIMENT

Machines turn data into knowledge. With every interaction they not only learn about the world but also about the person using them.

CHRISTOPH ENGEMANN AND
PAUL FEIGELFELD

After two extended periods of hibernation, artificial intelligence (hence-forth "AI") is with us once again. The first period of slumber was in the 1960s, after cybernetic laboratories had failed to bring their rule-based AI systems to the practical application stage, causing state and military financial backers to turn off the financial taps. The second wave of AI in the 1980s also ended in a funding freeze.[1]

After that, all went quiet again in the world of AI. Instead of lofty fantasies of computers achieving human-like feats and identifying complex relationships, science instead focused on gradually reproducing the mechanisms of biological systems. The body was at the centre of these endeavours, and instead of representing processes, such as walking, in an abstract form and then implementing them in robots, researchers worked from the outside in, so to speak. Simple leg constructions in which the individual joints were operated by a simple computer with a neural network of only two neurons proved much more effective than all previous attempts at teaching robots how to walk. This approach, typical of the nineties and noughties, was known as "embodiment".[2] Following the sudden appearance of highly manoeuvrable walking apparatus capable of competing in special high-profile soccer tournaments such as the RoboCup, developers argued that AI required a body in which it could reside if it were to exist at all. And, indeed, these apparatuses proved to be highly sensitive to the slightest changes in their bodies. They responded to changing degrees of friction in the joints, minimal torsion effects on the limbs, and varying foot structures with learning and adaptive strategies, including topical behaviour, such as the ability to seek out energy sources in their surroundings to satisfy their energy requirements. Hungry robotic bodies suddenly exhibited unexpectedly complex modes of behaviour. Intelligence, if we are to describe these functional spectrums as such, consists here of a combination of sensors, materials, and environments. It develops and rises to the challenge when it encounters physical resistance between its own body and the external world through which such devices are made to navigate.

In the latency period after the second hibernation of AI, people were convinced that if artificial intelligence were ever to become a reality, it could do so only in a form arising from embodiment.

253

1 Nils Nilson J., *The Quest for Artificial Intelligence* (Cambridge, Massachusetts, and London, Cambridge University Press, 2010), pp. 408f.
2 Andy Clark, *Being There: Putting Brain, Body, and World Together Again* (Cambridge, Massachusetts, A Bradford Book. The MIT Press, 1997); Andy Clark, *Supersizing the Mind: Embodiment, Action, and Cognitive Extension* (Oxford, Oxford University Press, 2008).

In the meantime, however, AI has emerged from its dormancy to greet the light of day once more. Its technical foundations are no longer bound by the embodiment approach, but by three factors that came together in around 2010: the availability of Big Data generated by social media and smartphones, the enormous increase in computing capacity thanks to special hardware such as graphics accelerators, and the opportunities these developments provide for advancements in the area of learning algorithms. The new volumes of data and computing capacities led to unexpected successes for long-existing algorithms designed for neuronal networks. In particular, deep learning, with its "recurrent neural networks" or deeply tiered, multi-layer neuronal networks, proved to be a mighty tool for transforming data into knowledge,[3] although it must be said that by knowledge we are referring less to abstract knowledge imbued with meaning and more to purely predictable behaviour. If one shows a recurrent neural network enough images of cats and lets it know that such images are described as "cats", then, after a sequence of training runs, it will identify pictures of cats as such with a high degree of probability.

These seemingly trivial achievements marked a new phase in the history of AI. Responding to the question of how the current boom in AI differed from previous incarnations, Demis Hassabis, cofounder of the company Deep Mind, replied: "This is the first time in history that AI has made money." Deep Mind is a British company founded in 2010 dedicated to researching AI. Before it was sold to Google for a rumoured 400 million dollars, a number of important international technology giants had already invested in the company, including PayPal cofounder Peter Thiel, Tesla founder Elon Musk, and the Hong Kong investor Li Ka-shing, one of Asia's most powerful men.

3 Pedro Domingos, *The Master Algorithm: How the Quest for the Ultimate Learning Machine Will Remake Our World* (New York, Basic Books, 2015).

Google's acquisition of Deep Mind was certainly one of the triggers that helped launch the current comprehensive restructuring of our technological world. Companies such as Facebook, Amazon, Microsoft, Intel, the Chinese Internet giant Baidu, and others quickly began purchasing entire academic teams before successively acquiring startups in the field of machine and deep learning beginning in the autumn of 2014. In the process, AI is being transformed from a technological niche into a comprehensive infrastructure. Every marker on a photograph appearing on Facebook and every Google search serves to help train neuronal networks, for which special

processors and other hardware are being produced on a massive scale. In September 2016, Facebook, Google, IBM, Intel, and Microsoft joined together to form what is known as the "Partnership on AI", which not only underscored AI's importance but also made critical problems more visible. Hassabis' statement, too, touches on this important moment in the history of AI: its relationship to economics, or more specifically, political economy, and the questions of where the financing for research and implementation actually comes from.

In the current wave of AI, politically motivated financing has moved aside to make way for economically based funding. The

wireless headphones recently unveiled by Apple are an example of this. Apple did not opt for wireless headphones in order to save space inside its latest iPhone. After the wrist, it wanted to install a computer on another part of the human body: the auditory region between the mouth and ear. This allowed Siri, Apple's AI assistant, to move into said space and profit from the greater numbers of interactions between user and device. The headphones contain four microphones, and instead of operating the graphical user interface, all a user must do is say the words "Hey Siri" in order to interact with the local device as well as the off-device Apple Cloud.

For the time being, then, machine learning is helping to lower the interaction threshold in three ways:

1.   The graphical user interface is becoming less important, and other, mainly auditory and haptic interface options – in other words, interactions relying on hearing and touch – are becoming more relevant. What these interfaces have in common is that they were both only made possible thanks to machine learning (henceforth "ML") processes, as user interactions are interpolated without the user having to perform symbolic interactions on the device. This lowers the level of friction involved using digital devices.

2.   The more interactions that take place, the more input is generated for ML models, enabling them to become ever more sophisticated.

3.   In this context it is important to understand the strategic motivation behind the evaluation of these interactions and their translation into economically utilisable analytics, whereby other information such as location data, time, level of activity, heart rate, and other factors are correlated as well.

Even more than Big Data, machine learning focuses on transforming the environment into an environment of transactions in which individual interactions in and of themselves probably are of little importance, but the sum of all interactions and their secondary and tertiary attendant circumstances can be monetised in advertising and insurance products.[4] In other words, deep learning is trying to reach ever deeper into our pockets and promises to expand the scope of these transactions. The financing of the current wave of AI development rests upon a very real dream: namely, the development of real counterparts that can be turned into money in all manner of contexts.

4 Christoph Engemann, "You cannot not transact. Big Data und Transaktionalität", in Ramón Reichert (ed.), *Big Data. Analysen zum digitalen Wandel von Wissen, Macht und Ökonomie* (Bielefeld, transcript, 2014), pp. 365–384; Florian Sprenger, Christoph Engemann, "Im Netz der Dinge", in Florian Sprenger, Christoph Engemann (eds.), *Internet der Dinge: Über smarte Objekte, intelligente Umgebungen und die technische Durchdringung der Welt* (Bielefeld, transcript, 2015), pp. 7–57.

These projects are being spearheaded by companies such as Facebook, Google, and Baidu, meaning that current developments in AI are being driven not by political blocs but by companies competing for market share – which they retain largely thanks to customer loyalty. And customer loyalty is set to increase as ML simplifies the use of digital devices and our interactions with them. Furthermore, the promise of automating processes that have thus far resisted all attempts at automation now seems to be just round the corner, driverless cars and trucks being the most well-known examples, while the automation of activities such as legal advice, medical diagnostics, or journalistic research and reporting seem not to be too far off either. The limits of machine learning and how the world of work, including creative jobs, will change remains unknown. The creativity of musicians, designers, and filmmakers is already being called into question by compositions, 3D-printed objects, and films dreamt up by deep neuronal networks.

It is important to point out that over the course of its development the term "machine learning" increasingly has given way to the term "artificial intelligence". This is both unfortunate and telling, for what is lost is the notion of "learning". As many authors and scientists in the field of deep learning have commented, the notion of learning represents a fundamental step forward vis-à-vis the old approach. In contrast to the technology of the first two waves of AI, deep learning does not depend on formalised and deductive processes with explicitly formulated rules. Instead, ML involves inductive or abductive processes that rely not on formal rules but on data training and the formation of behavioural probabilities based thereon.[5] As described above, after having processed a couple of hundred thousand images, deep convolutional networks react to images of cats by producing the label "cat": to paraphrase Nietzsche, these networks are superficial – out of profundity.

257

5 Peter Norvig et al., *Artifcial Intelligence: A Modern Approach* (New York, Pearson, 2009); Pedro Domingos, *The Master Algorithm: How the Quest for the Ultimate Learning Machine Will Remake Our World* (New York, Basic Books, 2015).

Pedro Domingos – a researcher in the field of ML and author of the admittedly rather sensationally titled book *The Master Algorithm*[6] – believes the transition from programming to learning represents a fundamental discontinuity in the evolution of the computer. No longer will people function as the cognitive agents who bring the world to the computer. Instead, computers will learn about the world all on their own, and in doing so will come to recognise significantly more complex relationships than humans will ever be capable of doing with their limited capacity for processing information.

What Domingos ignores, however, is the role of humans as the signalling selectors in these learning processes, for the most successful ML processes embrace the notion of "supervised learning". In these types of ML processes, a training set labelled by humans – such as photographs of cats or house numbers – serves to train a network, which is then confronted with unlabelled images and is tasked with assigning them the correct label. The output then must be classified by humans. Notably, both the training and the output classification are carried out using real people by means of crowdsourcing. As Oren Etzioni, CEO of the Allen Institute for Artificial Intelligence (financed by Microsoft cofounder Paul Allen), said of the process, "Deep learning is still ninety-nine percent human work."

Thus we can say that we humans serve as the environment for the training of ML processes. This form of AI is based upon embodiment by proxy: the machine's lack of a body has to be compensated for using human agents. One could see the human body as an impulse generator and part of the system's recursive materiality. This makes ML's above-mentioned occupation of the body by means of microphones in headsets or pulse monitors on the wrist not only possible but also necessary. ML is capable of learning from things that humans themselves often miss or cannot explain.

This is the thread that runs all through the history of learning: the differentiation between formal and informal learning and, simultaneously, that between institutional and lifeworld-based learning.[7] Those involved in learning and education have sometimes seen opportunities for emancipation in one, sometimes in the other. Currently, tacit knowledge[8] is all the rage in the humanities as the type of knowledge capable of negotiating the supposed intractability of non-symbolic and hence formalised knowledge that is the prerogative of educational institutions.

6 Pedro Domingos, *The Master Algorithm* (see note 5).
7 Anna Tuschling, Christoph Engemann, "From Education to Lifelong Learning: the Emerging Regime of Learning in the European Union", in *Educational Philosophy and Theory,* (vol.38, no.4, 2006), pp. 451–469.
8 Micheal Polanyi, *Implizites Wissen* (Frankfurt a. Main, Suhrkamp, 1985).

9 Nick Bostrom, *Superintelligence: Paths, Dangers, Strategies* (Oxford, Oxford University Press, 2014); Benjamin H. Bratton, "Outing Artificial Intelligence: Reckoning with Turing Tests", in Matteo Pasquinelli (ed.), *Alleys of Your Mind: Augmented Intelligence and Its Traumas* (Lüneburg, Meson Press, 2016), pp. 69–80.

Christoph Engemann studied psychology in Bremen and wrote his doctoral dissertation in media studies at the Bauhaus University Weimar on the electronic identity card. This was followed by research fellowships at the University of Texas at Austin, the Oxford Internet Institute, and at the International College for Cultural Technology Research and Media Philosophy at the Bauhaus University Weimar. At present, Engemann is participating in a postdoctoral programme in the German Research Foundation's (DFG) research group Media Cultures of Computer Simulation (MECS) at the Leuphana University of Lüneburg. His research interests include media transformation of the state, digital authentication media and their history, governmediality, and rurality and barns.

Now that the economic dimension of ML revolves around generating ever more money-making opportunities for interaction, it would seem logical to describe the above process as the mechanisation of tacit knowledge: the humans in possession of such knowledge remain unaware of the fact, but for the operators of ML processes such knowledge is transformed into a dimension that makes its evaluation at least theoretically possible – "theoretically" being the key word, for the internal representations of convolutional neural networks are still not adequately understood and continue to elude mathematical representation.

For people, learning is always simultaneously a limiting necessity and a door to self-determination. After all, learning provides a basis for taking our own decisions, and it is precisely this self-determination that has been regarded as the specific privilege of humans ever since the Enlightenment. But the arrival of ML poses a new form of knowledge circulation between people and organisations that aggregates in the Cloud, where new decision-making processes that are often incomprehensible for humans evolve. The really interesting question that the discourse surrounding AI derives from this development and carries forward is: Are computers capable of learning their way out of their determined state into self-determination?[9]

Paul Feigelfeld is a graduate in cultural studies and computer science. From 2004 to 2011 he worked for Friedrich Kittler, one of Germany's most notable media theorists, and also co-edited Kittler's collected works. From 2010 to 2013 he was a lecturer and researcher in the Institute for Media Theory at the Humboldt-University in Berlin before becoming academic coordinator of the Digital Cultures Research Lab at the Centre for Digital Cultures of the Leuphana University of Lüneburg. He is currently working on his dissertation "The Great Loop Forward. Incompleteness and Media between China and the West". He regularly advises museums and festivals and writes for publications such as *frieze, Texte zur Kunst,* and *PIN-UP.*

For the time being, AI remains dependent on the bodies of others. Without bodies integrated into the processes of machine learning that can permanently generate signals and thus create a topology of learning opportunities, ML will remain dumb, so to speak. ML delegates this embodiment to us, and as a distributed embodiment it draws on a sensory environment made up of mobile phones, wireless headsets, smart watches and smart homes, networked automobiles, and a wide variety of other devices. ML's lack of a body will gradually cease to be an issue, and a new AI will arise out of the concurrence of sensors, materials, and environments made possible by Big Data and smartphones in which the human body functions merely as an impulse-generating transit station. It remains to be seen just who will be whose pet in the future.

AF

260

WOULD YOU
A ROBOT?

DO YOU
WANT
TO
BECOME
BETTER
THAN
NATURE
INTEND-
ED?

LIVE IN

Architecture with sensors ...

Philip Beesley. *Hylozoic Soil,* Musée des beaux-arts de Montreal, Quebec, 2007

... that reacts to its environment.

Philip Beesley's "living architecture" blurs the line between nature and technology and imagines architecture as a living, layered system capable of interchange with its users. It aims to create a "metabolic architecture" which conceives of manmade structures not as inanimate objects, but as living systems capable of evolution and growth, and envisions a new, more immersive architectural paradigm. Since 2008, the initiative has been developed through the *Hyozolic* series, a set of immersive installations that use motion-tracking and touch sensors to react and adapt to the movements of people passing through them. OP

PHILIP BEESLEY – *HYLOZOIC GROVE*

Philip Beesley. *Hylozoic Grove,* 2016. Acrylic, mylar, borosilicate glass, memory shape alloy; installation ca. 200 × 500 cm, columns ca. 240 × 30 × 30 cm each, filter clusters 15 / 30 × 60 × 60 cm each © Philip Beesley Architect Inc. (PBAI)

ASMBLD – *PROJECT DOM INDOORS*

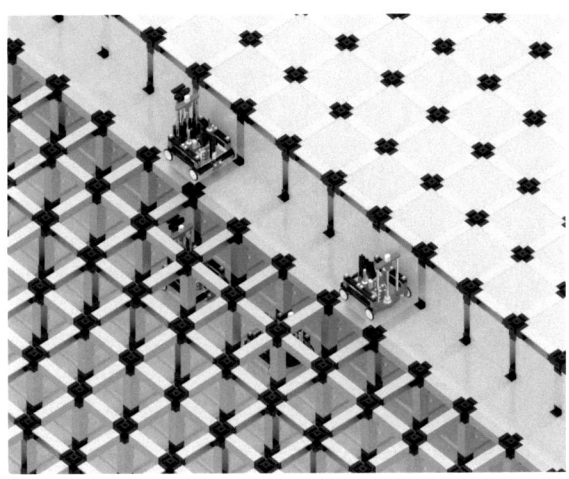

Project Dom Indoors, 2015

Asmbld. *Project Dom Indoors,* 2015.
Robotically reconfigurable interiors,
rendering © Asmbld Architectural
Robotics

In times of rising rents – particularly in dense urban
areas – new strategies must be found for making flexible
and effective use of our ever-dwindling living and working spaces. *Project Dom Indoors* offers a possible solution:
using robots capable of assembling and disassembling
standardised architectural construction elements, interior
spaces can be reconfigured to support more flexible uses.
For example, robots can quickly raise interior walls, ped-
estals for mattresses or sofas, and desks and kitchen
counters. Beginning at floor level, the robots build up
   and take down – the structures layer for layer, allowing
for the most diverse spatial configurations. Although
the system does not yet conform to our modern understanding of homeyness, it is still possible to imagine a
variety of possible uses and contexts. TT

Greg Lynn presents us with a vision of future habitats with his *Room Vehicle Prototype,* a cocoon-like form constructed at a scale of 1:5 that automatically rotates 360 degrees horizontally and 180 degrees vertically. This allows the entire interior surface to be used for various functions depending on the angle of inclination – the occupant simply has to move along with the rotating interior. A total area of sixty square metres of living area is "wrapped" around the interior, saving both space and resources compared to living spaces with conventional floors, walls, and ceilings. Lynn was inspired by space stations and robot-controlled lounge chairs that provide every amenity from refreshments to entertainment within arm's reach. Although chairs like these are often associated with passive consumption, Lynn's rotating residential units – for which he has been known to invoke the image of the hamster wheel – are designed as habitats for active people. TT

Greg Lynn. Scale Model for *RV Prototype,* 2013. 1/20 model attached to motorised base, carbon fiber twill with epoxy resin, fiberglass, 3D printed plastic, stepper motors, timing belts, computer script, and electronics, 63.5 × 43.2 × 45.7 cm, courtesy San Francisco Museum of Modern Art, gift of the artist © Greg Lynn

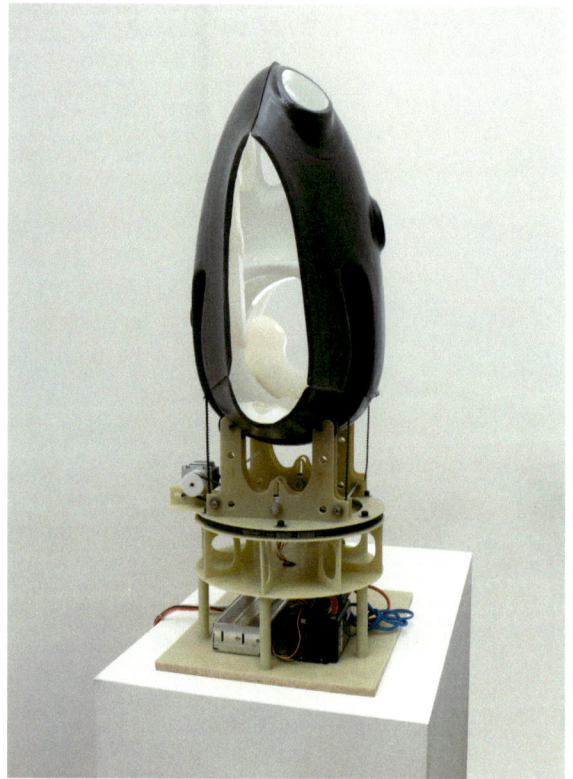

Scale model for *RV Prototype,* 2013

GREG LYNN – *RV PROTOTYPE*

HyperHabitat: Reprogramming the World, 2008

Guallart Architects, IaaC, MIT The Center for Bits and Atoms, FAB LAB Network and Bestiario. *Hyper-Habitat: Reprogramming the World,* 2008. Interactive installation, acrylic, LEDs, internet 0 node, various sizes; directed by Vicente Guallart; Guallart Architects: María Díaz; Institute for Advanced Architecture of Catalonia: Daniel Ibañez, Rodrigo Rubio, Marta Male Alemany, Areti Markopoulou, Laia Pifarré, Alessio Carta, Christian Zorzen, Vagia Pantou, Daniel Bas, Stefania Sini, Francisca Aroso, Melissa Mazik, Maria Papaloizou, Luis Fernando Odiaga, Georgios Machairas, Ismini Koronidi, Anastasia Fragoudi, Ifigenia Arvaniti, Georgia Voudouri, Hemant Purohit, Renu Gupta, Luciano Bertoldi, Peerapong Suntinanond, Javier Olmeda, Raya Alexandra Theodorou, Higinio Llames, Susana Tesconi´, Nuria Sanz, Panagiota Papachristodoulou, Luis Casado Martinez (electrician); MIT The Center for Bits and Atoms: Neil Gershenfeld; FAB LAB Network: Victor Viña, Tomas Diez, Lucas Cappelli; Bestiario; Andrés Ortiz, Santiago Ortiz, José Aguirre, Daniel Aguilar; partners: Visoren, Irpen, Luz Negra, Ministerio de Vivienda Ajuntament de Barcelona; communication: Pati Nuñez © 2017 Institute for Advanced Architecture of Catalonia, photo: Jose Morraja

GUALLART ARCHITECTS / IAAC / CENTER FOR BITS AND ATOMS MIT / BESTIARIO – *HYPERHABITAT: REPROGRAMMING THE WORLD*

The installation *Hyperhabitat: Reprogramming the World* originates from research done at the Institute for Advanced Architecture of Catalonia (Iaac) on the interconnectivity of physical spaces inspired by features of existing digital networks such as nodes, protocols, line codes, and so on. By giving real objects digital identities, the project aims to restructure our existing living networks – not only within our homes, but also between different neighborhoods, cities or even continents – saving energy and making us more efficient. For the original installation at the 12th Venice Biennale, a representative "home" (in this case comparable to a floor of a student dormitory) was filled with over thirty laser-cut objects made of methacrylates resembling furniture, appliances, and even a crucifix. Each was embedded with micro-severs and thus able to communicate with other objects via a fictive platform. EP

265

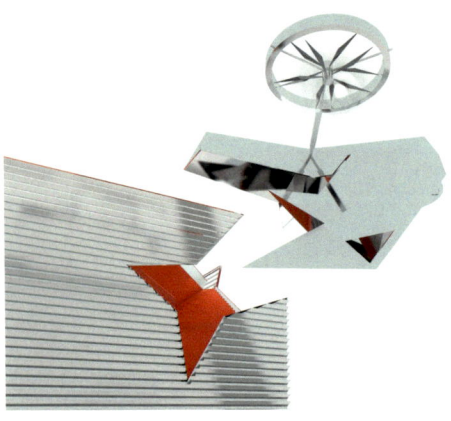

Nestled Crystals, 2014: Here, form ...

... follows movement.

Nestled Crystals is a result of a collaboration between Suprastudio at the University of California's architecture department and the aerospace giant Boeing. This collaboration sparked students to rethink the mobility and dynamics of buildings in light of robotics technology. Ismael Soto's work envisioned a flying, movable hotel tower that can dock and nest in different parts of the building through the use of electric rotors. The rotor blades are equipped with LED projectors that can show images when in motion, thereby functioning as a large display. The idea is that the flying object could be used as a billboard and as an event location. The ambiguity between the soft motion and the hard edges of the static cube creates an air of urban dynamism. AR

ISMAEL SOTO – *NESTLED CRYSTALS*

Ismael Soto. *Nestled Crystals,* Mondrian Hotel Redesign, West Hollywood, California, 2014. Advisors: Greg Lynn, Julia Koerner & Boeing; architectural model, 27 × 27 × 17 cm © Ismael Soto

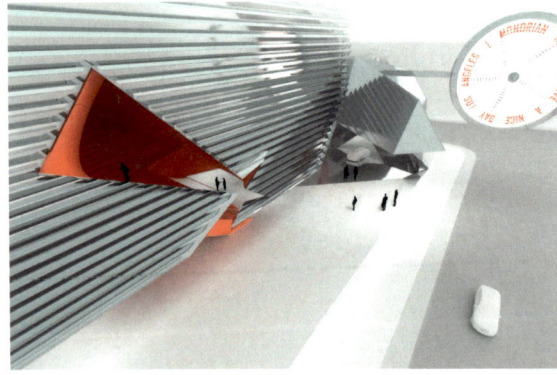

Exterior view

Interior view

HÖWELER + YOON ARCHITECTURE, SQUARED DESIGN LAB – *ECO PODS*

Eco Pods is a concept for sustainable urban energy production. The *Eco Pods* are modular, gondola-like containers that can be attached to one another or to existing architecture. The pods function as bio-reactors housing populations of micro-algae that provide a hyper-efficient source of bio fuel. Large robot arms, powered by the energy generated, constantly rearrange the algae pods to provide optimum lighting and growing conditions. The *Eco Pods* are designed to function simultaneously as bio-power stations, research laboratories, and means of communicating the potential of this sustainable energy source to the public at large. TT

Höweler + Yoon Architecture, Squared Design Lab. *Eco Pods,* 2009. Rendering; design team: Franco Vairani (Squared Design Lab) and Daniel Sullivan © Höweler + Yoon Architecture, LLP

Eco Pods, 2009: An algae farm...

... that's always moving.

GRAMAZIO KOHLER RESEARCH & RAFFAELLO D'ANDREA / ETH – *FLIGHT ASSEMBLED ARCHITECTURE*

Residential architecture built one step at a time by flying robots and without the touch of a human hand or a single construction crane – this is the vision presented by architects Gramazio Kohler Research and roboticist Raffaello D'Andrea in their installation. Like a swarm of buzzing bees, a number of four-propeller helicopters stack building elements according to mathematical algorithms. A digital design is translated into the behaviour of the flying machines, causing them to cooperate and construct a building together. The installation is a 1:100 scale model for an architectural vision that envisages a 600 m "vertical village" for 30,000 residents in Meuse near Paris. TT

Vertical Village

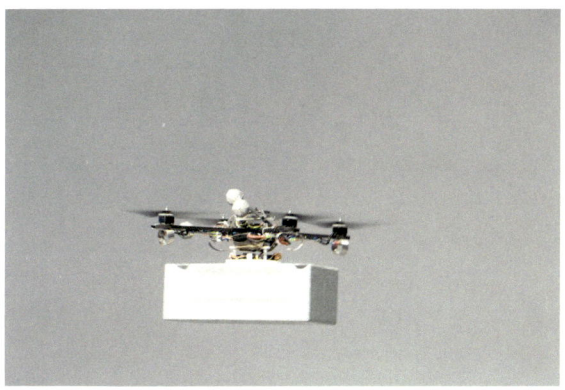

... built by drones... © photo: François Lauginie

Gramazio Kohler Research & Raffaello D'Andrea in cooperation with
ETH Zurich. *Flight Assembled Architecture,* FRAC Centre Orléans,
2011–2012. Architectural installation with four quadrocopters and
model © Gramazio Kohler Research,
ETH Zurich

A visualisation of the different floors in the *Vertical Village*

The project is located on the TGV train route between Strasbourg and Paris

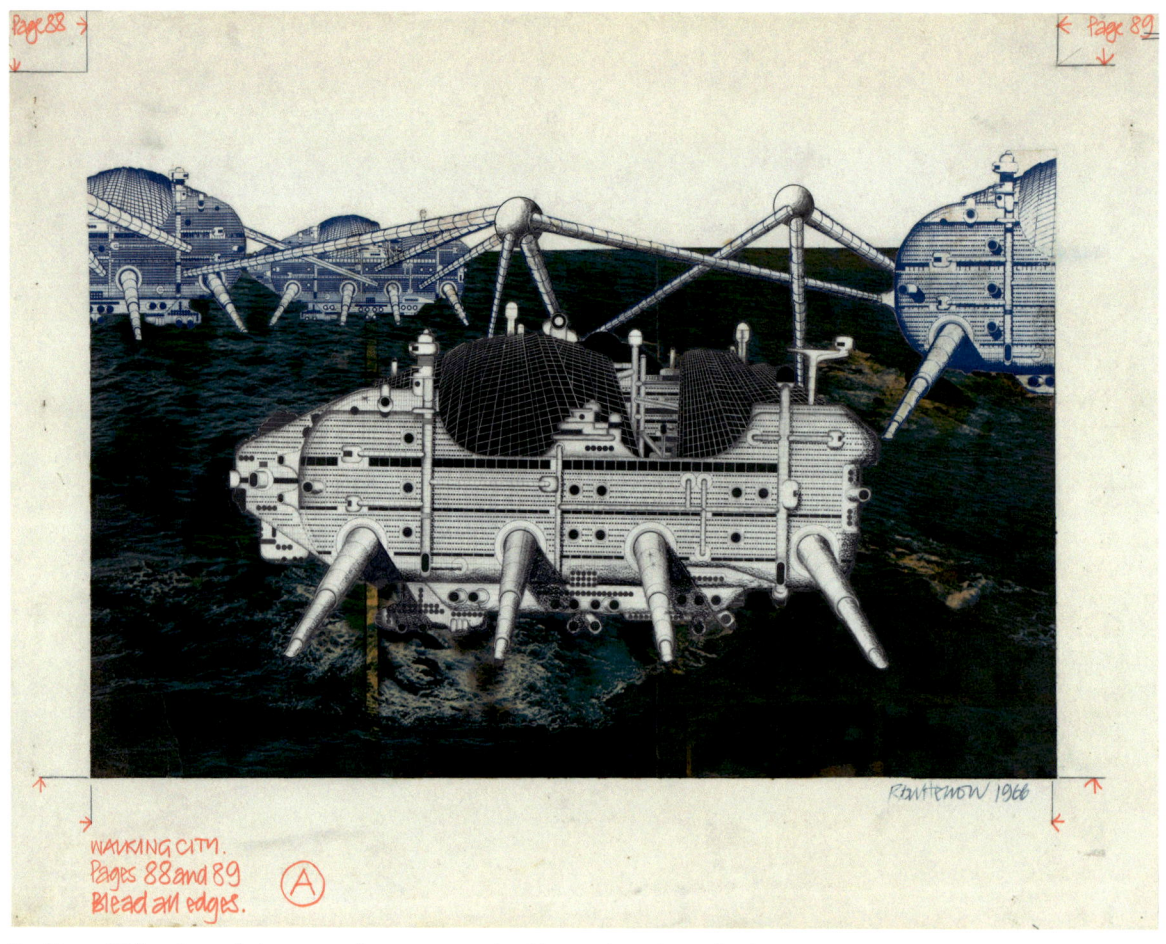

Ron Herron. *Walking City on the Ocean, project (exterior perspective),* 1966. Cut-and-pasted printed and photographic papers and graphite covered with polymer sheet, 29.2 × 43.2 cm; gift of The Howard Gilman Foundation, acc. no.: 1203.2000 © 2016, digital image, The Museum of Modern Art, New York/Scala, Florence

Ron Herron. *Walking City: Elevations of Vehicle,* 1964. Facsimile with original drawing by the artist, 102 × 72 cm (framed); courtesy Archigram Archival Project, the University of Westminster © The Ron Herron Archive London

"A house is a machine for living in" is a well-known aphorism coined by the architect Le Corbusier in 1921. British architect Ron Herron – a member of the avant-garde group Archigram – took the idea of such a "living machine" literally with his 1964 design *Walking City*. He developed massive container-like robotic structures whose mobility and intelligence allowed them to go wherever they were needed. They could temporarily cluster together into cities and then disperse again. Yet this vision of futuristic nomadism was set in a rather dystopian context: Herron was imagining life after a nuclear catastrophe. LH

RON HERRON, ARCHIGRAM
– WALKING CITY

INTERACTIVE ARCHITECTURE LAB, UCL – *REEARTH: HORTUM MACHINA, B*

The autonomous robotic ecosystem *Hortum machina, B* is half garden and half machine. A spherical structure inspired by Buckminster Fuller's geodesic architecture contains twelve garden modules with native plants. These modules can be extended individually in order to shift the structure's centre of gravity, allowing the sphere to slowly move forward. The rules according to which it does so are derived from the electrophysiology of the plants themselves and their collective "intelligence", since plants react electrochemically to their environment. These collective signals are translated into the sphere's control system, which, depending on lighting or sources of noise, causes the sphere to move. The garden machine thus represents a concept for urban gardening in the cyber-age, which can also contribute to spreading native seeds. TT

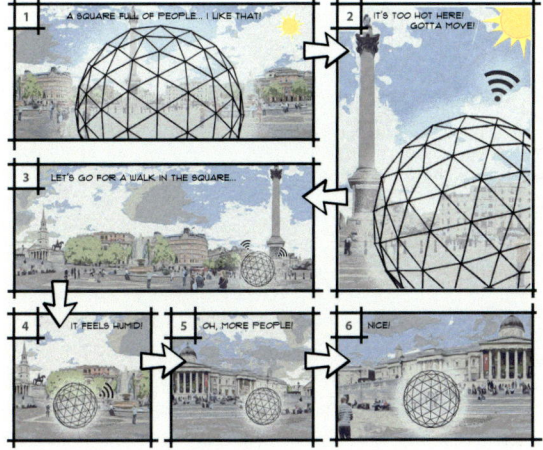

reEarth: Hortum machina, B, 2016

Interactive Architecture Lab, UCL. *reEarth: Hortum machina, B,* 2016. Digital drawing; team: William Victor Camilleri, Danilo Sampaio; director IALab: Ruairi Glynn; engineering & robotics: Christopher Leung, William Bondin, Francesis Mangion and Thomas Powell © Interactive Architecture Lab, Bartlett School of Architecture, UCL London

UNIVERSAL EVERYTHING – *WALKING CITY*

Universal Everything's 2014 video *Walking City* was not only inspired by the iconic *Walking City* series (made by the 1960s architecture group Archigram's member Ron Herron) – the video has also become a kind of twenty-first century successor to its namesake. Matt Pyke, the founder of Universal Everything, teamed up with animator Chris Perry and sound designer Simon Pyke to create the video, in which a walking figure seamlessly transforms into a series of architecturally-inspired forms varying in abstraction. Thanks to a symbiosis of fluid computer-generated animation and a hypnotizing beat, the audiovisual composition is mesmerizing for the entirety of its duration, just under eight minutes. It reminisces about the mobility and autonomy of Herron's clunky, roaming robo-ships while looking forward to a more flexible, adaptive kind of robot organism. *Walking City* won the Golden Nica at the Ars Electronica in 2014 and has been exhibited worldwide. EP

Universal Everything. *Walking City,* 2014. Video, 7 min 47 sec; creative director: Matt Pyke, animation: Chris Perry, sound: Simon Pyke © Universal Everything

Walking City, 2014: The roaming city assumes every possible form ...

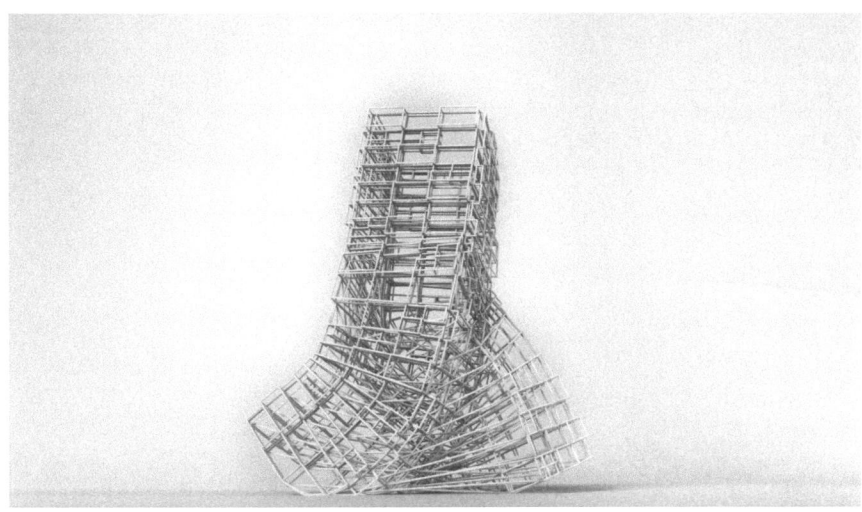

... every possible structure ...

... and every possible materiality

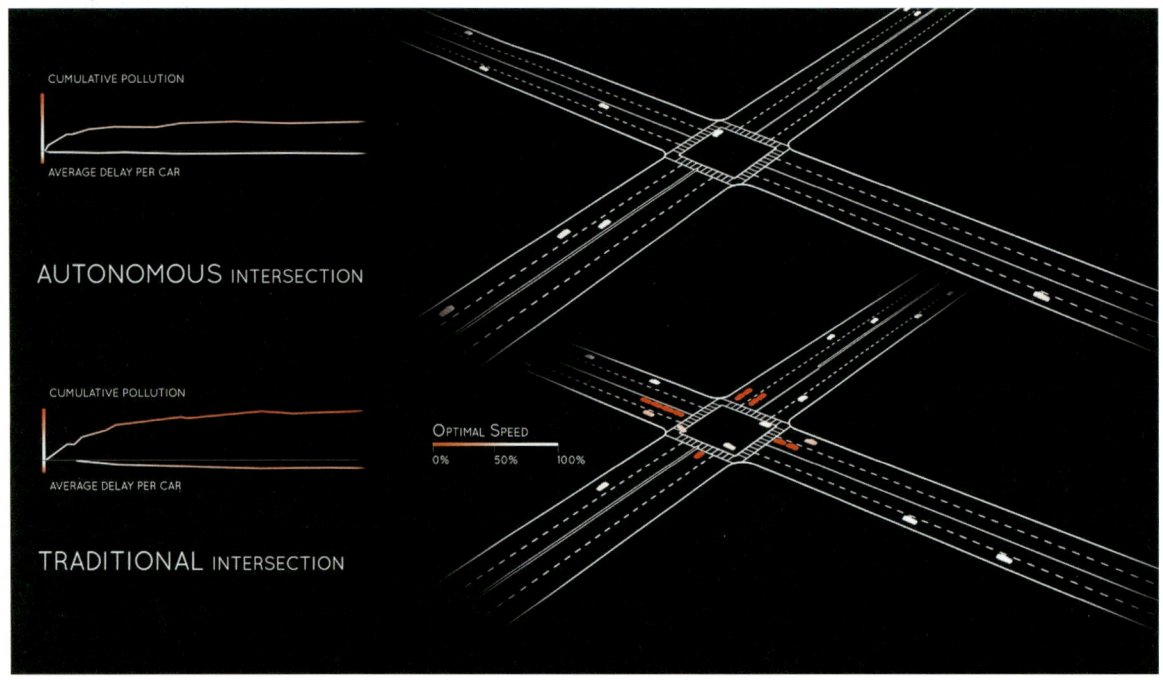

DriveWAVE, 2014

MIT SENSE*ABLE* CITY LAB – *DRIVEWAVE*

MIT Sense*able* City Lab. *DriveWAVE,* 2014. Video, 1 min 26 sec © MIT Sense*able* City Lab

In a future world in which self-driving cars are a given – as has been predicted for the year 2030 – will we still need traffic lights to make us stop and wait at intersections on our way home from work? No, say the researchers at the MIT Sense*able* City Lab. Their *DriveWAVE* is a digital traffic control system, a "smart intersection" which can calculate gaps in traffic at lightning speed and guide the networked vehicles through the intersection without stopping. It is fast enough to ensure a steady flow of traffic while still allowing for sufficiently safe distances between the individual vehicles. This will not only enable us to get from A to B more quickly, but will also cut fuel consumption by eliminating the constant need for vehicles to brake and accelerate. According to the researchers' model for the future, Traffic 4.0 will be fluid and seamless. TT

Richard Vijgen. *Architecture of Radio,*
2015. App; photo: Juuke Schoorl
© 2016 Richard Vijgen Studio

Created by Richard Vijgen, *Architecture of Radio*
is a data visualisation app that displays the invisible system of data cables, radio signals, cell
towers, and satellites that we depend on for
communication, observation and navigation.
Based on the user's GPS location, the app shows
a 360 degree visualisation of the signals in the
area and aims to provide a comprehensive win-
dow into the infosphere. The app borrows from
global open data sets of cell towers, Wi-Fi routers,
and observation satellites, revealing the invisible
technological landscape we interact with
through our devices. OP

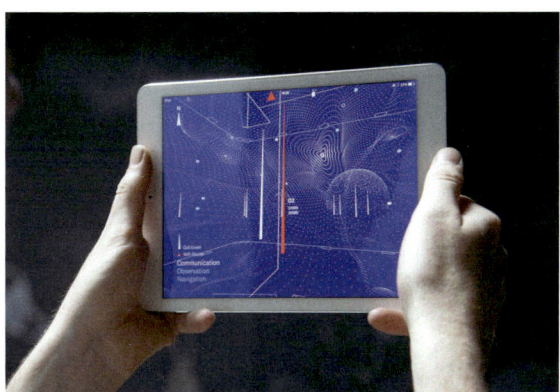

Architecture of Radio, 2015

RICHARD VIJGEN – *ARCHITECTURE OF RADIO*

WOULD YOU L

DO YOU WANT TO
TER THAN NATUF

A
A
E

N A ROBOT?

BECOME BET-
INTENDED?

ROBOTS
NCING
UTION?

BRUNO BIANCHI, ANDY HEYWARD, AND JEAN CHALOPIN – *INSPECTOR GADGET*

Bruno Bianchi, Andy Hayward, and Jean Chalopin. *Inspector Gadget,* 1983–1986. Animated cartoon series, 86 episodes © "INSPECTOR GADGET", "GO, GADGET, GO!", "GO GO GADGET!" and all related titles, logos and characters are trademarks of DHX Media (Toronto) Ltd.

Inspector Gadget is a cartoon series that originally aired from 1983 to 1986, but continues to be aired in syndicated television reruns across the world to this day. The series was co-produced by companies in France, Canada, and the USA. Every episode follows the adventures of the bumbling cyborg detective Inspector Gadget, as he fights the forces of "M.A.D.", the evil organisation run by Dr. Claw. In his battles, he uses an arsenal of gadgets built into his body, such as pop-up roller-skates, mechanical arms, and a helicopter hat. Although he is well equipped, all his tricks tend to backfire. If the Inspector ever succeeds in his quests, it is thanks to the help of his niece Penny and her dog. AR

Inspector Gadget, 1983–1986

278

DANGEROUS THINGS LLC
– *XBTI*

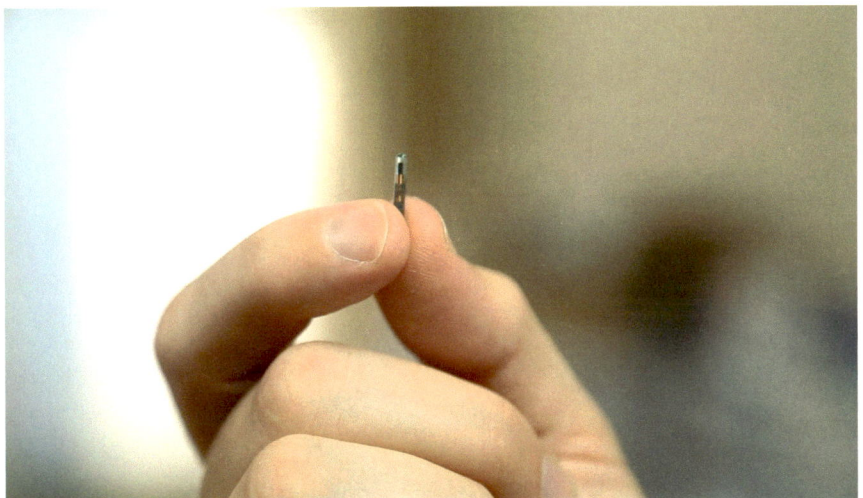

xBT-Chip, 2016

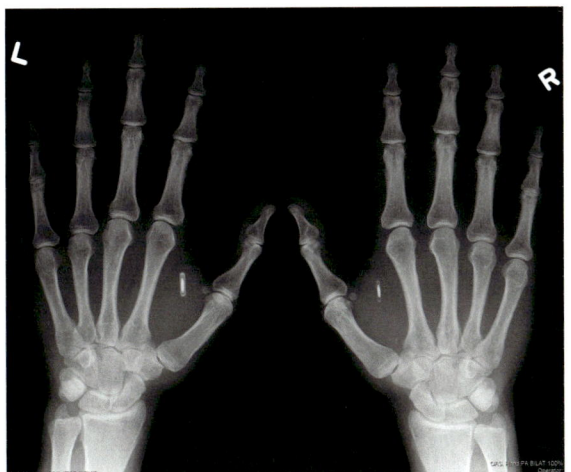

X-ray photograph with an implanted *xBT* chip, 2016

Dangerous Things. *xBTi* [xBT + injection kit], 2016. Implantation set with chip, ca. 10 × 1 × 1 mm
© Dangerous Things

xBT is a chip developed for insertion under the skin of the human hand – the same technology that has become widespread for identifying and labelling animals. The unique ID number stored on the chip can be used as a code for a variety of functions: for example, keyless entry to house or car doors simply by touching them with the hand and unlocking cell phones without needing to remember or write down a PIN code. Products such as the *xBTi* set (including chip and sterile implant package), which is already available on the open market, pose the question of how far we are willing to go with modifications and optimisations to our own bodies – procedures known as bodyhacking – in the service of digitalisation and networking. TT

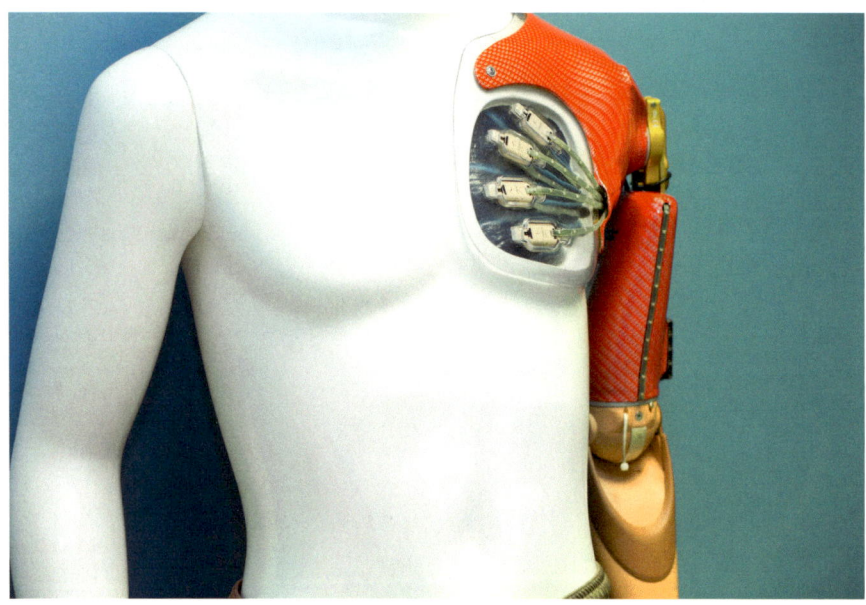

Dynamic Arm Plus, (ongoing)

Ottobock in collaboration with the
AKH Vienna / Faculty of Medicine,
University of Vienna. *Dynamic Arm
Plus,* ongoing. Intuitively controlled
prosthetic arm; TMR mannequin,
glass plate, 190 × 58–60 × 30 cm
© Ottobock, photo: Christian Apostol

OTTOBOCK – *DYNAMIC ARM PLUS*

In 2007 the medical technology company Ottobock successfully introduced thought-controlled prosthetic hands and arms. To use them, patients who have lost an arm initially have the remaining nerve endings reactivated. This creates an interface to the human brain on the surface of parts of the shoulder and chest. Nerve impulses sent from the brain can be relayed via these interactive "human-machine interfaces" to the motorised prosthesis. With a little practice, the wearer can then mentally control their reconstructed bionic arm and get it to perform ordinary everyday movements.

Aktion Mensch. *Neue Nähe*, 2016.
Video, 5 min 4 sec; (music) title:
"Aufklärung", composers: Tim van
Berkestijn, Michiel Marsman,
publisher: Sizzer Music, courtesy
Tempomedia © Aktion Mensch e. V.,
photo: Henrik Hühnken, Zacharias
Zitounie

Neue Nähe, 2016

AKTION MENSCH – *NEUE NÄHE (NEW CLOSENESS)*

"Every new idea can bring us closer together": this is the slogan of the film from the Aktion Mensch 2016 promotional campaign. In the film, which had already received 1.4 million clicks on YouTube only six months after being posted, able-bodied children encounter adults and children with disabilities. The latter use the most modern prosthetics, such as exoskeletons or bionic hands, as well as hi-tech devices like speech processors and iPhones that help them to speak and see. Seen through the eyes of a child, there's nothing too futuristically strange about these technologies – they're simply exciting. Reservations are quickly dispelled, and it is not the disability that commands attention, but rather the special technology and all the things that can be done with it. LH

The *Ekso GT*™ is the first licensed motor powered robotic exoskeleton to help patients whose lower extremities have been paralysed after a spinal cord injury or a stroke to stand up and learn to walk again. Steps are triggered by the patient's shifting weight laterally. The training programme can be accurately tailored to suit individual needs in real-time. Ekso calculates and adjusts the path of each step through its swing phase at a rate of 500 times per second, thus more effectively mimicking an unimpaired gait. Walking aids to support the arms such as crutches, walking frame, arm slings, and a walking cane are also included, and depending on patient progression these are used in different stages of the rehabilitation process. During a therapy session, the software provides the therapist with real-time information about the interaction between patient and machine and thereby allows for the analysis and documentation of the patient's achievements. LH

EKSO BIONICS – *EKSO GT*™ *ROBOTIC EXOSKELETON*

Ekso Bionics®. *Ekso GT*™*Exoskeleton,* 2013. Robotic exoskeleton, suitable for people between 150 and 190 cm tall, net weight: 27 kg, self-supporting © courtesy Ekso Bionics® Europe GmbH

Ekso GT™*Exoskeleton,* 2013

Leka, 2016: Game ...

LEKA SAS – *LEKA*

Leka SAS. *Leka,* 2016. Pedagogical robot for children with disabilities, 30 × 22 × 22 cm © Leka SAS

... and therapy, used with autistic children.

Leka is a robot product, a "smart toy", developed to interact with autistic children who are usually prone to withdraw into worlds of their own, as well as to help parents and therapists to communicate with them. A smart, spherical, portable toy, it reacts to voices, touch, and movement and can play hide-and-seek. Its purpose is to involve children and help them train their motor, cognitive, and sensory skills. Leka's behaviour is predictable enough to give the children confidence, but its colours, sounds, and vibrations can be configured individually in order, for example, to prevent overstimulation. The first trials have shown that robots can be useful in situations of insufficient interpersonal competence. TT

WAFAA BILAL – *3RDI*

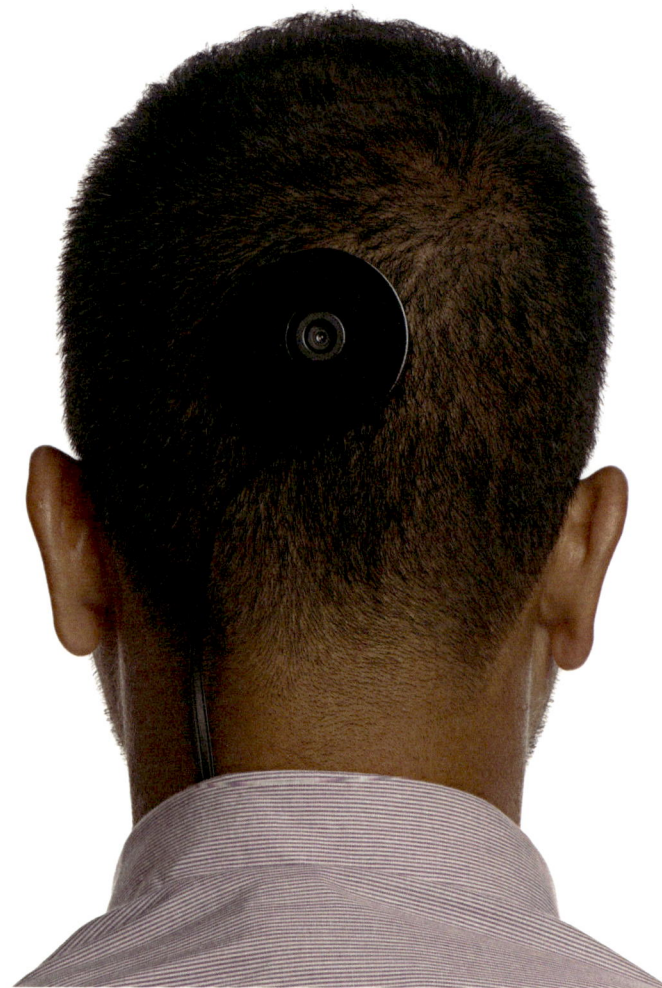

3rdi, 2010 – 2011

Wafaa Bilal. *3rdi,* 2010–2011.
Photo installation, inkjet prints on
luster paper © Wafaa Bilal, courtesy
Driscoll Babcock Galleries and
Mathaf: Arab Museum of Modern Art

In 2010, Iraqi artist Wafaa Bilal had a webcam surgically implanted onto the back of his head
to capture random images of his everyday life for a whole year at a rate of one snapshot per
minute, 24 hours a day. The images were streamed live to Bilal's website, which thus displayed
what the artist – quite literally – had left behind. Bilal called the project "anti-photography"
since he could not see or select the images being captured. Using his own body as a medium,
he offered a platform to discuss the pervasive, but sometimes unnoticeable, technological surveillance that is part of our daily lives. AR

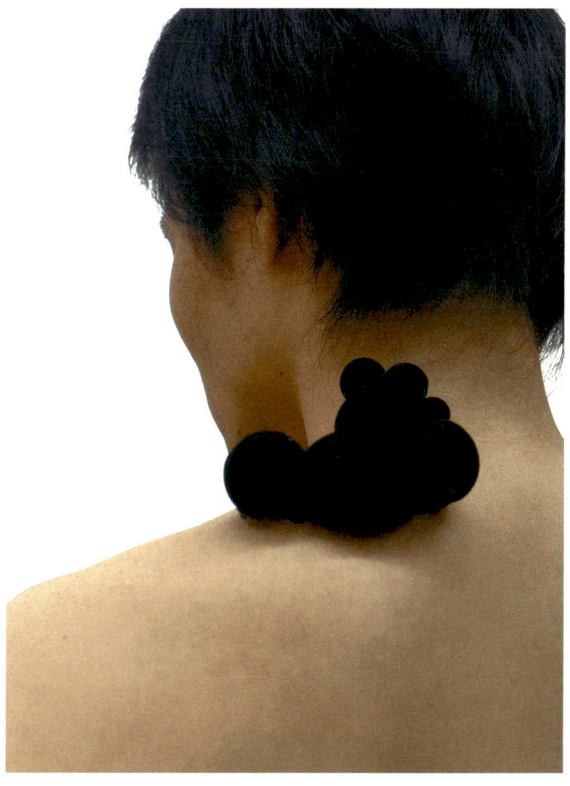

Prostheses for Instincts – Form Prototypes, 2009

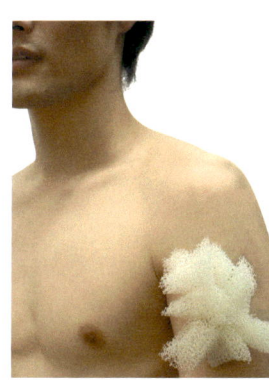

Prostheses on the skin ...

Behind Susanna Hertrich's half-artistic, half-scientific experimental project lies the idea of technically expanding the natural human sensorium to enlarge the perceptual spectrum and enable us to sense other factors such as risks. A variety of prostheses devised by the artist are worn directly on the skin and communicate streams of data that correspond to sources of danger, stock exchange information, natural disasters, or local crime rates. These in turn create sensations in the wearer that are familiar to us from our natural warning instincts: a cold shiver, goose pimples, etc. The question underlying the project is whether and how machines can be made to produce human emotions. TT

SUSANNA HERTRICH – *PROSTHESES FOR INSTINCTS – FORM PROTOTYPES*

Susanna Hertrich. *Prostheses for Instincts – Form Prototypes,* 2009. Photographs, 20 × 15 cm each, mixed media objects: 15 × 15 cm each © Susanna Hertrich

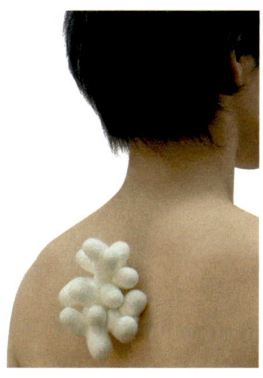

... forewarn of modern perils, ...

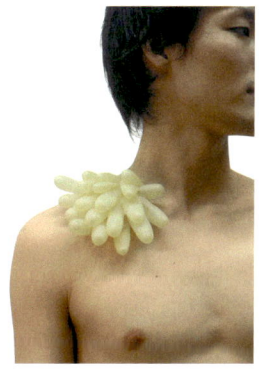

... such as a stock market crash.

DO YOU
WANT TO
BECOME
BETTER
THAN
NATURE
INTENDED?

ARE RO
ADVANC
EVOLUT

WOULD YOU
LIVE IN
A ROBOT?

OTS
NG
ON?

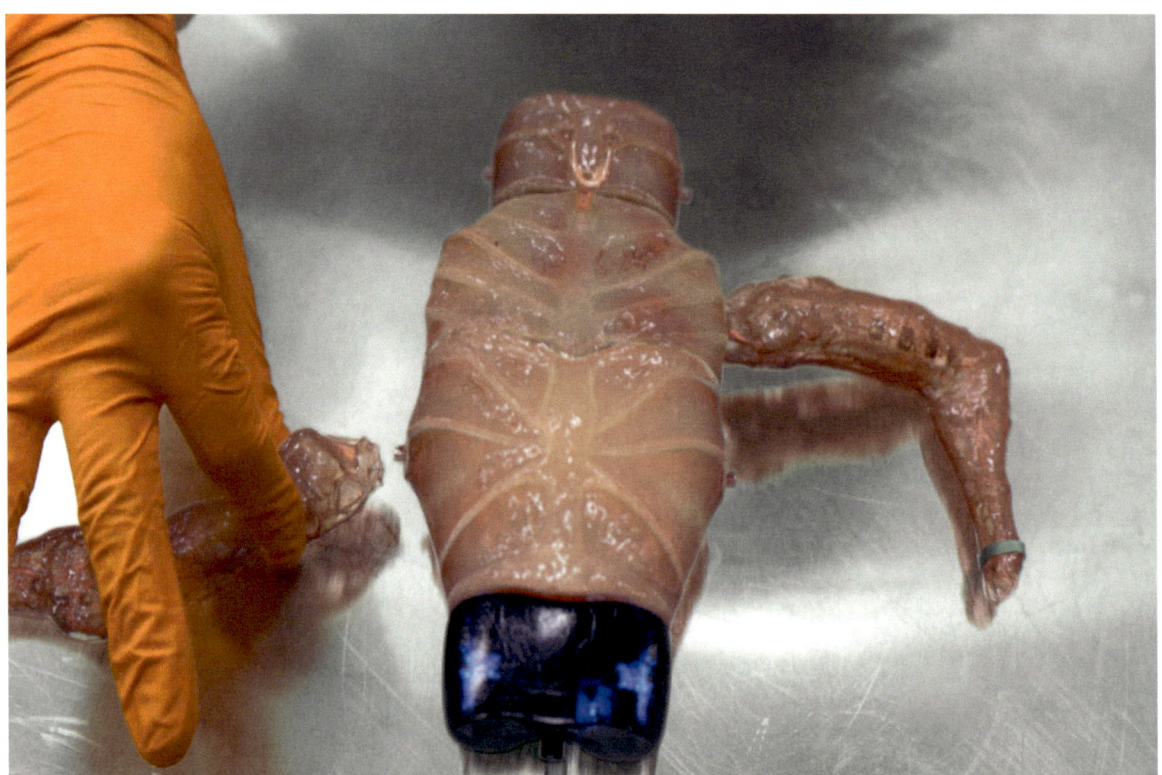

Oscar: The Modular Body, 2016

Oscar is the first prototype of a copy of the human organism composed of a number of clickable modules that function independently of one another and are hence capable of independent survival. Or so Dutch artist and film director Floris Kaayk would have us believe when he presents this *Gesamtkunstwerk* on his website – and it takes a while to see through his very realistic staging. For his Homunculus project, Kaayk demonstrated his love of gory detail in a video; here, Oscar's ostensible inventor Cornelis Vlasman not only presents *Oscar: The Modular Body,* but also stages made-up media reports and blog contributions about him. LH

FLORIS KAAYK – *OSCAR: THE MODULAR BODY*

Floris Kaayk. Oscar: *The Modular Body,* 2016. Web art project composed of 56 videos; concept, direction: Floris Kaayk; research, script: Floris Kaayk, Ine Poppe; interface design: LUSTlab; animation: Floris Kaayk, Adriaan van Veldhuizen; online media strategy: Nienke Huitenga; production: seriousFilm; co-production: VPRO © 2016 seriousFilm, VPRO

Oscar's homepage

NewBees, 2014

Greenpeace. *NewBees,* 2014. Video,
2 min 22 sec; director: Polynoid,
creative director: Alexander Kalchev
© Greenpeace, courtesy DBB Paris

The video maps out a future scenario in the wake of a worldwide pollination crisis. Bee populations are extinct and pollination is no longer possible – with catastrophic consequences for agriculture and biodiversity.
Advanced technology has come to the rescue, developing a new generation
of bees, the *NewBees.* These solar-powered robot bees have been introduced around the world and are efficiently doing their duty, pollinating
plants and efficiently eliminating enemies. They are easy to handle, require
only brief charging periods, and are highly durable. The scenario allows
Greenpeace to pose – and answer – the critical question of whether pre-
serving the existing natural world might not be a more attractive option
and a better guarantor of our continued existence than non-stop technical
innovation. TT

GREENPEACE – *NEWBEES*

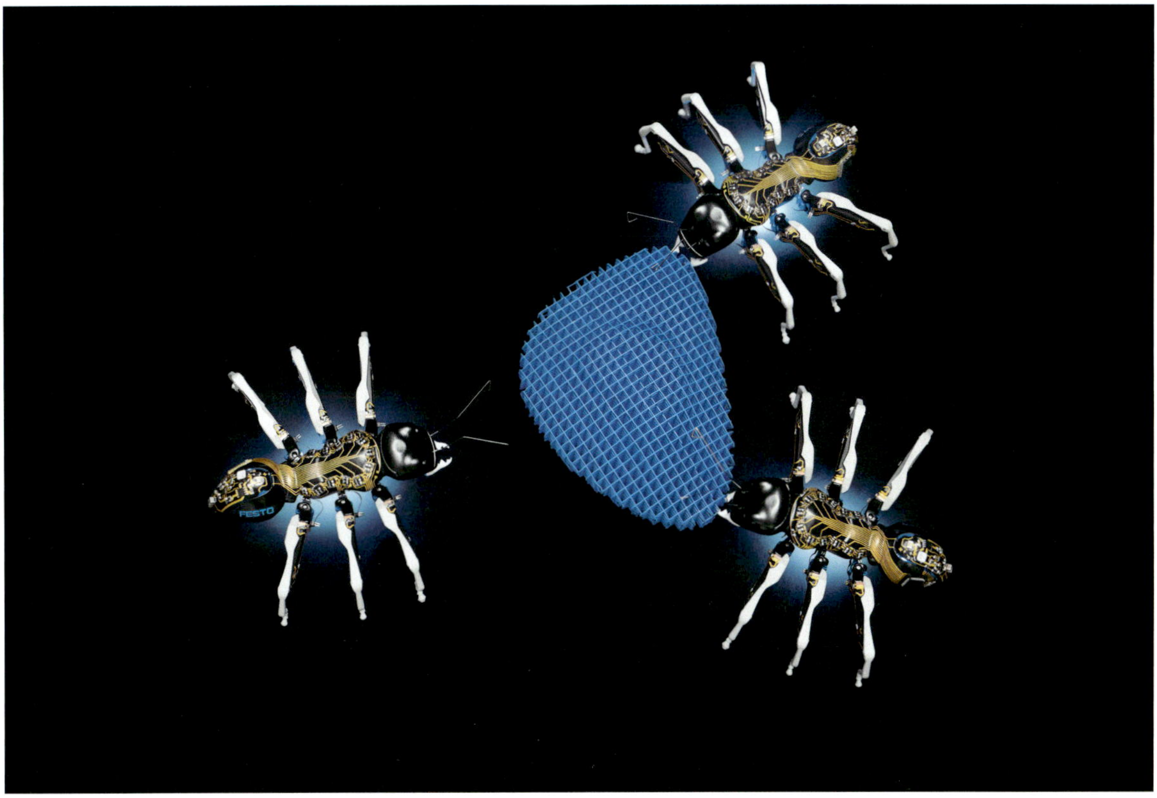

BionicANTs, 2015

BionicANTs not only replicate the delicate anatomy of real ants, they also transfer their cooperative behaviour into the world of technology with the aid of complex control algorithms. Although the robotic ants are much larger than their natural role models, they still manage to pack a great number of components and sensors into a very small space. They communicate with one another and coordinate their actions and movements. The goal of this research and development project is to create artificial intelligence that can arrive at complex solutions thanks to the co-operative behaviour of many individual sub-systems – behaviour that is analogous to that of ants. Industry 4.0 will require machines capable of close networking and cooperation in order to facilitate flexible production. TT

Cooperative robotic ants

Festo. *BionicANTs,* 2015. Robotic ants, laser-sintered polyamide and 3D MID laser structuring, 4.3 × 13.5 × 15 cm each © 2017 Festo AG & Co. KG

FESTO – *BIONICANTS*

ICD/ITKE, UNIVERSITY OF STUTTGART – *RESEARCH PAVILION 2013/14*

Research Pavilion, 2013/14

Since 2010, the Institute for Computational Design (ICD) and the Institute of Building Structures and Structural Design (ITKE) at the University of Stuttgart have created an annual Research Pavilion. Both institutes share a distinctive focus on biomimicry and material experimentation, designing structures that are robust yet singular. In the 2013–14 academic year, a team of researchers drew their inspiration from beetles and, in particular, from their lightweight protective wing-cases known as elytra. These derive their stability from the intricately wound fibres of which they are composed, while their light weight is due to a central cavity. Following this model, the team programmed two collaborative robots to wind glass and carbon fibres together to create modular panels. Additionally, thanks to the algorithm, it was ultimately the robots rather than the architect who decided how and where the individual panels would be placed. EP

A robot winds together glass and carbon fibres ...

ICD/ITKE, University of Stuttgart. *Research Pavilion,* 2013/14. Pavilion, glass and carbon fibres, 50m², 12kg/m² basis weight. Architectural model, 1:20 scale, various materials, 75 × 90 × 50 cm © ICD/ITKE, University of Stuttgart

... according to the computations of an algorithm.

ANOUK WIPPRECHT
– *SPIDER DRESS 2.0*

Anouk Wipprecht's creations herald the end of analogue clothing. She regards fashion as an interface and combines design with robotics and electrical engineering, integrating microchips and sensors into her fabrics and into the structures of her designs, which she makes with a 3D printer. The "smart clothing" she produces using these techniques – so far mainly one-off conceptual prototypes – functions like an intelligent second skin. The multi-jointed, movable arms on the collar of her *Spider Dress 2.0*, for instance, register the speed with which someone is approaching and if necessary will reach out to mark the wearer's personal space. LH

Anouk Wipprecht. *Spider Dress 2.0,* 2015. Robotic dress, various materials, 3D printed, with Intel Edison microcontrollers © Anouk Wipprecht, photo: Jason Perry

Spider Dress 2.0, 2015

Molecule Shoe, 2015

FRANCIS BITONTI STUDIO – *MOLECULE SHOE*

Francis Bitonti's *Molecule Shoe* is a 3D-printed collection of pixelated platform footwear, developed by an algorithm that mimics cellular forms in nature. The shoes were created with a mathematical model called "Game of Life", which generates cellular structures. This algorithm allowed the designer to "grow" pairs of shoes with varying gradients of colour; each shoe has a slightly different form. Once the form is created, the shoes are built pixel by pixel on a Stratasys 3D printer that blends different colours of filament while building up the pieces layer by layer. OP

Francis Bitonti Studio Inc. *Molecule Shoe,* 2015. 3D printed with the Stratasys Connex 3D Printer, printing software by Adobe; pair of shoes, 23.5 × 24.1 × 10.2 cm each; photo © 2017 Museum of Fine Arts, Boston

ICD, University of Stuttgart. *Aggregate Architecture,* 2015. Column, ca. 200 × 50 × 50 cm © ICD, University of Stuttgart

Aggregate Architecture, 2015

For the project *Aggregate Architecture* at the ICD at the University of Stuttgart, researchers investigated the possibilities of aggregates – large amounts of elements in loose contact – in architecture. The researchers developed their own synthetic granular systems based on naturally forming granulates like sand and gravel. The new granules are produced from recycled plastic in various sizes, depending on what best suits the specific structure being built. For construction, small clusters are dropped into position by a cable robot based on an algorithm. Ultimately, the goal is to develop a system with granules so small that no gaps can be detected. Since no matrix is needed for construction, this building technique would be ideal for temporary purposes and quick reconfigurations. EP

ICD, UNIVERSITY OF STUTTGART – *AGGREGATE ARCHITECTURE*

For his constantly expanding art project *Slogans for the Twenty-First Century,* the Canadian author and artist Douglas Coupland collects slogans and thoughts with which he tries to understand the ways in which the present – in the relatively early days of the twenty-first century – is fundamentally different from the twentieth century. It comes as no surprise that this collection of ideas primarily focuses on life with and in the midst of digital media, which has become second nature in this day and age. Slogans such as "Machines are talking about you behind your back", "You are the last generation that will die", and "Offline = Loneliness" can seem ominous or unsettling and socio-analytic by turns. Taken as a whole, the collection constitutes a kind of oracle for life in the digital age. TT

DOUGLAS COUPLAND – *SLOGANS FOR THE TWENTY-FIRST CENTURY*

Douglas Coupland. *Slogans for the Twenty-First Century,* 2011 (ongoing). Pigment prints on watercolour paper, laminated onto aluminium, 35 pieces, 55.4 × 43 cm each © Douglas Coupland, courtesy Daniel Faria Gallery, Toronto

GETTING SHITFACED IN A DRIVERLESS CAR IS GOING TO BE AWESOME

THE UNANTICIPATED SIDE EFFECTS OF TECHNOLOGY DICTATE THE FUTURE

TECHNOLOGY FAVOURS HORRIBLE PEOPLE

MACHINES WILL MAKE BETTER CHOICES THAN HUMANS

THEY REALLY WILL

KILLING PEOPLE WITH DRONES IS CHEATING

USE JETS WHILE YOU STILL CAN

FRIENDS DON'T UNFRIEND FRIENDS

UNCREATIVE PEOPLE USE THE WORD 'ROBOT' AS A TOOL TO FREAK YOU OUT

MACHINES ARE TALKING ABOUT YOU BEHIND YOUR BACK

WHAT IF THERE'S NO NEXT BIG THING?

HACK YOUR DRIVERLESS CARS TO DESTROY THE SPEED LIMIT

PEOPLE MAKE BAD CHOICES WHEN TECHNOLOGY CHANGES TOO QUICKLY

IT'S ALL HAPPENING WAY FASTER THAN WE THOUGHT

TALKING ABOUT ROBOTS MEANS YOU HATE CHILDREN AND NATURE

LOOKING BACKWARDS WON'T HELP THIS TIME

ROBOTS ATE YOUR JOB FOR BREAKFAST

ROBOTS PUSH YOUR CLASS BUTTONS

A LOT OF PEOPLE DON'T WANT PROGRESS

HEALTHY PEOPLE ARE BAD FOR CAPITALISM

MACHINES WILL MAKE BETTER CHOICES THAN HUMANS

EXCESS LEISURE TIME IS A DISASTER

ROBOTS DON'T BUY FURNITURE

YOU. ME. DRIVERLESS CAR. TONIGHT.

SCIENCE FICTION IS NOW JUST FICTION

DRONES AND ROBOTS ARE BEST FRIENDS

TECHNOLOGY IS THE ONLY REMAINING LINK TO ENLIGHTENMENT

PERMANENT REVOLUTION

OH... I SEE YOU'RE NOT USING A MAC...

ROBOTS YEARN TO DETHRONE YOU

MACHINES ARE TALKING ABOUT YOU BEHIND YOUR BACK

HOARD ANYTHING YOU CAN'T DOWNLOAD

IN THE FUTURE WE'LL ALL BE SHOPPING FROM JAIL

HONK IF YOU'RE DRIVING A DRIVERLESS CAR

DRONES YEARN TO SEE YOU NAKED

THE PAST IS NOW USELESS

WELCOME TO DETROIT THE WHOLE WORLD IS NOW DETROIT

Slogans for the Twenty-First Century, 2011 (ongoing)

GLOSSARY

3D-Print
See rapid prototyping

Act(uat)or
An act(uat)or or effector is a machine element which triggers movements, for example the grasping movement of a robot hand or the rotary motion of a robotic vacuum cleaner's wheels etc. Often they are engines or pneumatic or hydraulic cylinders or other kinds of actuators. Besides **sensors** and calculators, actuators are the main components of robots.

Algorithm
An algorithm is a sequence of orders that can help to solve a task. Algorithms work like recipes, issuing orders like take this and do that until a certain purpose is achieved. They are to be found everywhere in our daily lives: converted into software, they find the best routes for navigation, search results or suggestions. Robots are equipped with various algorithm-based skills.

Android
See humanoid

Artificial Intelligence
Artificial intelligence (AI) refers to machines that, like humans, are capable of intelligent behaviour, meaning they can think logically, use knowledge, plan, learn, process language, and perceive the world. Recently, social intelligence and creativity have also begun to play a role in AI. A transdisciplinary field of research combining computer science, mathematics, psychology, linguistics, neuroscience, and other fields, AI seeks to describe human intelligence in terms detailed enough to allow it to be formalised and simulated using computer programs; other approaches attempt to analyse and reconstruct the information architecture of the human brain with the aid of neural networks. One of the greatest obstacles to AI is that we do not know how human intelligence actually works.

Augmented Reality
Augmented reality (AR) or mixed reality is the perception of reality which is expanded by the help of computers – often, but not always via visual illustrations. In contrast to the **virtual reality**, which aims for an experience caused by entirely computer-generated surroundings, AR works with pictures or videos which are superimposed with additional, computer-generated information or virtual objects. Commonly known examples of AR are the digital input of working lines at football broadcasts, diverse layers in touristic maps or computer games in open landscapes.

Automat
An automat (derived from the Latin *automatus,* "voluntarily acting of one's own accord") is a machine, which executes predefined processes independently, thus "automatically". Like a robot, an automat works in a self-contained manner: when triggered, it performs a mechanical process without human help (e. g. vending machines).

Automation
Automation is the transmission of human work to automats as technology progresses. Since the early abaci and water- and windmills, automation has increased – from the invention of engines and sophisticated mechanisms that spurred the industrial revolution all the way to the computer technology of the twentieth and twenty-first centuries. In the face of increasing robotisation and its impact on the labour market, discussions have begun about alternatives to gainful employment such as the unconditional basic income. Given the upsurge of robotization and its impact on the labour market, today there are even discussions about alternatives to gainful employment such as the unconditional basic income.

B

Big Data

The term Big Data describes the large volume of often fast moving and exponentially growing digital data in our networked world. This data comes from sources like customer databases, digital communication, route guidance systems, social networks, search engines etc. Big Data also denotes the ambition to utilise the gathered information – mostly for commercial and intelligence-gathering purposes. The collection and analysis of data has become a daily routine in the fields of market- and trend research, financial management, climate simulation, technological development, and many other fields. As Big Data applications use personal information, they raise major privacy and data security issues.

Biohacking

Biohacking refers to the intervention in biological processes by amateur biologists; in particular, the practice of a group of activists who use freely available technologies to perform **trans-human** changes on themselves or on others. In order to endow the human body with capabilities beyond its natural ones (keyword: human enhancement), computer chip implants are inserted, and magnets, **sensors,** or measuring devices are placed under the skin. In addition, genetic engineering is used. The conceptual boundary between bodies changed via biohacking and **cyborgs** is fluid. Currently still a niche, the connections between bodies and technologies are constantly increasing.

Bionics

The concept, a portmanteau of the words "biology" and "electronics", deals with the transfer of natural phenomena to the field of technology. Structures and processes that have been tried and optimised over millions of years of evolution provide ideas for innovative applications and answers to technical issues. These range from animal bodies and plant growth to the behaviour of slime mould.

Bot

A bot (from "robot") is a semi-autonomous agent in computer systems and the Internet designed to fulfil a specific task. In some instances, it is even capable of learning. Bots include chatbots – with which users can converse (one historical example being Joseph Weizenbaum's ELIZA, while modern examples include customer service bots or assistants such as Apple's Siri or Amazon's Alexa) – and Twitterbots that compose their own Tweets on specific subjects for advertising or political ends. A network of computers or **IoT** devices infected with viruses or malware is known as a "botnet". Botnets can carry out extensive cyber attacks, such as DDoS (Distributed Denial of Service) attacks, which can bring down larger systems or infrastructures (e.g., electricity or telecommunication networks), or even the entire Internet in a given country.

Botnet

See bot

C

Chatbot
See bot

Cyberspace
The term Cyberspace (from cybernetic space) is sometimes used to describe the Internet as a whole. More precisely, however, it refers to a three-dimensional, virtual world created through computer programmes. Coined by science fiction author William Gibson, the term is a combination of the word "space" and a derivative of "cybernetics". Cyberspace can refer to an environment of experience, operation, and work represented on the computer monitor as well as to a computer-generated space in which we completely immerse ourselves just like in **virtual reality**.

Cyborg
The short form for "cybernetic organism" defines a hybrid creature formed from living, especially human, organisms and machines. While human machines with supernatural powers are frequently found in science fiction literature, the practical application is more about optimising the body with artificial elements such as high-tech prostheses (also see **biohacking**).

D

Deep Learning
The method of deep learning is applied to teach machines a certain, specialised way of thinking. A form of machine learning, it is complex, requires a lot of computing capacity, and orients itself on the functioning of the human brain. Using artificial, multi-layered (thus "deep") neuronal networks, many calculations are made simultaneously. In this way, robots acquire the ability to identify things in images (a driverless car can, for example, detect pedestrians), and to recognise and process language, including various dialects, among many other things. Thinking or intelligence is limited to very small areas here and is not part of general intelligence.

Design Fiction
See speculative design

Drone
A drone is an unmanned aerial vehicle (UAV), which is either navigated via an on-board computer or by remote control (fully or partly autonomously). Drones, whose measurements range from micro-drones of a few centimetres to the size of commercial aircraft, are used for military, intelligence, civil and commercial but also scientific purposes. In addition, they serve as carriers of weapons and measuring devices, suppliers of goods to otherwise inaccessible areas or conflict zones and for reconnaissance, assessment, and documentation.

E

Exoskeleton
See prosthetics

Humanoid

A humanoid or humanoid robot is a robot whose shape fundamentally resembles the human body and is thus anthropomorphic. It does not necessarily have to look like a real human being, but it generally has a torso, limbs, and a head. This may be for technical-functional reasons, for example, if the robot is supposed to use human tools, or for social ones in the case of interaction with humans. An **android,** on the other hand – although the word is often used as a synonym – is a robot designed to resemble humans as closely as possible. Female versions are sometimes referred to as gynoids.

Industrie 4.0

Industrie 4.0 stands for the current revolutions in the industrial environment brought about by networked information and communication technologies. Following the mechanisation that started in the late eighteenth century, the subsequent electrification and mass production through assembly lines, and **automation** with **industrial robots** in the twentieth century (Digital Revolution), we have now reached the era of digital networking also known as the fourth industrial revolution. Based on a strategy paper by the German Federal Government to promote the computerisation of production, the term Industrie 4.0 ("Industry 4.0") is also widely used. Machines, robots, logistics, and products are now supposed to communicate and collaborate directly with each other, which should ultimately lead to production that is largely self-organised, i.e., "smart factories".

Industrial Robot

See automation

Interaction Design

Interaction design creates interfaces between humans and machines. In the late 1980s, as our communications environment became more and more complex, especially due to graphic user interfaces, a design discipline specialising in this field was conceived. Interaction design focuses on user-centred design and optimised user experience, be it in the designing of mobile phone displays, the interaction with robots that understand language, or the development of service processes. In view of our "increasingly intelligent" environment, interaction design is of growing importance.

Interface

The interface is the part of a system that enables communication. In communication between humans and machines, the interface is the point at which the user interacts with the device. This can be a switch or a computer's graphical user interface, or, in interaction with robots, their facial expressions, speech function, or the manipulator arm with which industrial robots assist us in our factories.

Internet of Things

The Internet of Things (IoT) refers to the growing number of digitally networked devices, vehicles, buildings, and other objects and thus, according to this definition, the information society's physical infrastructure. Equipped with miniaturised computers as well as **sensors** and **actuators**, these networked objects are set to play an ever-greater supportive role in our everyday lives, even if we are sometimes unaware of it. Concrete examples are **smart home** applications or a printer that orders its own ink online when the printer is nearly out of ink. Experts estimate that the IoT will comprise some 50 billion objects by the year 2020. A number of massive security risks inherent to the IoT came to light in 2016 when it was discovered that many of these objects are insufficiently secured, allowing them to be misused for botnet attacks.

M
Mechatronics
See robotics

P
Prosthetics

Prosthetics is the science that develops prosthetic devices – in other words, artificial substitutes for lost organs and body parts or extensions thereof. As a discipline that has long since combined the fields of **bionics** and mechanics – and later, electronics – it is of great relevance for robotics. For example, high-tech prostheses are now capable of performing the human hand's complex functions. If we can think beyond human models as the standard for prostheses' performance spectrum, it is also possible to envisage applications with non- or superhuman capabilities (see "**cyborg**" and "**biohacking**"). Even today, athletes with artificial legs are capable of greater speeds than athletes with natural limbs.

Prosumer

A term coined by the futurist Alvin Toffler in the 1980s, "prosumer" combines the words "producer" and "consumer". It refers to a person who both creates and uses products or services. The term is of particular relevance today within the context of user-generated content on the Internet and social networks. In the Web 2.0, content is produced by the same people who consume it. With the free availability of **3D printers** and robots as well as open-source construction manuals, the production of a wide variety of goods by prosumers will be possible.

R

Rapid Prototyping

The term refers to techniques for the rapid manufacture of sample components. Here, CAD (computer-aided design) data is implemented quickly and without additional manual work. Rapid prototyping, which can also take the form of "rapid manufacturing" (in which special finished parts instead of models are produced), generally makes use of primary forming techniques in which the work piece is constructed from shapeless material layer for layer, as in **3D printing**.

Robotics

A scientific discipline, robotics is devoted to the development of robots. Robotics also draws on branches of other disciplines, such as mathematics, electrical engineering, and computer science. More recently these areas have been combined to create the field known as mechatronics, which is of particular importance in the development of robots. The development of ever-more autonomous systems requires input from a growing number of disciplines, such as neuro-informatics and **bionics**. The term "robotics" first appeared in 1942 in the short story "Runaround" by the biochemist and science fiction author Isaac Asimov, in which the author lays out his "Three Laws of Robotics". These established ethical standards for the discipline.

S

Self-driving Vehicle

A self-driving vehicle is an automobile or airplane that is capable of smoothly functioning in traffic without the need for a human operator. Fully autonomous models with the necessary **sensors** and steering mechanism can also be described as "robotic cars". Whereas self-driving vehicles only exist today as prototypes, a number of IT and automobile companies are working on bringing the technology to the production phase. A number of ethical questions and control issues still need to be resolved. Who is responsible if accidents occur? What decisions should the control mechanism make (should it cause the driver's or a pedestrian's death in an emergency)? Can the vehicle be hacked and piloted remotely?

Sensor

A sensor (from the Latin *sentire*, meaning "to feel") is a technical component capable of registering particular physical or chemical characteristics in its environment and transforming these into electrical signals. Sensors play an important role as signal generators in automated processes and in robotics. They recognise values that can then be processed by a control system and trigger subsequent actions or reactions. A robot can therefore react "intelligently" and change the direction of its movement if it, for example, registers the existence of a wall in front of it using ultrasound sensors. Sensors are found everywhere in our surroundings, from the flushing mechanisms in our toilets to smoke and movement detectors.

Singularity

The term singularity has been used by a number of influential thinkers (John von Neumann, Hans Moravec, Ray Kurzweil) since the 1950s to describe an event predicted to result from technical evolution. This event marks the moment at which technology catches up and finally overtakes humanity at both the physical and the mental levels. The singularity theory is tightly interwoven with **artificial intelligence** and **transhumanism**. Both utopian and dystopian future scenarios are based on the concept.

Smart City

The term Smart City was coined in the 2000s to refer to a concept in urban development intended to make cities more technologically progressive, sustainable, efficient, and inclusive. It involves technological, economic, and social innovations, with a key role assigned to digital technologies and networking – for example, in the planning of energy-efficient and low-emission mobility, in the introduction of regional circular economies with minimal transport distances, or increasing participation, the sharing culture, and E-democracy.

Smart Device

Smart, connected products or smart devices are products or devices equipped with computers, sensors, software, and network connectivity, which enables them to communicate and exchange information with their surroundings as well as with other devices, users, and products. This communication is often external and done by means of a cloud service. Taken as a whole they make up the **Internet of Things** and play a role in **ubiquitous computing**. Examples of smart devices are autonomous robotic vacuum cleaners or smart home applications in which household devices and home utilities are automated and networked with one another, allowing them to be controlled and / or programmed from anywhere.

Smart Home

See smart device

Social Robot

Social robotics develops autonomous machines, usually humanoid in appearance, which adhere to social rules, display social and emotional behaviour, and can interact with humans. The great challenge faced by this young sub-discipline of robotics is ensuring that the robots interpret human behaviour correctly and can respond adequately in line with their assigned roles. In practice, artificial social intelligence has many potential areas of operation: in the service industry, in the care and nursing services, or within the context of **Industry 4.0**, in which there is a great demand for close cooperation between humans and machines.

Speculative Design

As a practice and a strategy speculative design anticipates desirable applied product innovation scenarios. In more general terms, it is a kind of design research and a strategy that seeks to address intangible phenomena and issues that are difficult to grasp as a way to spur the imagination and visualise possible processes of innovation between society and technology. The term was introduced by the British designers Anthony Dunne and Fiona Raby as an expansion of their concept of critical design, a term they also coined.

U

UAV – unmanned aerial vehicle
See drone

Ubiquitous Computing

The trend in which the PC yields to more mobile, networked computers embedded in a variety of devices is referred to as ubiquitous computing. Following the first era of large mainframe computers and the second era of the PC – a computer for everyone – experts are now talking about ubiquitous computing as the third era of the computer in which a large number of objects and terminals in the human environment come together to form a network (see also **Internet of Things**). Embedded in these devices are small and mobile networked computers, and in many instances **sensors** and **actuators**. The term "ubiquitous robotics" is virtually synonymous with ubiquitous computing.

Uncanny Valley

The phenomenon, also called "acceptance gap", refers to an effect in our acceptance of artificial human replicas. According to this effect, first described by the Japanese roboticist Masahiro Mori in 1970, people are less willing to accept robots that resemble humans and find abstract, completely artificial figures more attractive and acceptable. Only when the imitation is perfect does acceptance begin to rise again.

T

Transhumanism

The term stands for a school of thought that seeks to expand human potential by means of technology. Transhumanists think the human body should also be subject to technological progress. They believe genetic engineering, brain-computer interfaces, high-tech **prosthetics**, and the development of superintelligence will offer people a better quality of life. Related schools of thought are to be found in the **biohacking** community and in the figure of the **cyborg**. If we apply the theory to history, the wooden leg and eyeglasses could also be considered transhumanist.

V

Virtual Reality

The perception of reality in a computer-generated, virtual environment in real time is referred to as virtual reality. The aim of this phenomenon is to completely immerse the user in a virtual world, whereby the subject's perception of his- or herself in the real world is reduced. This is different from **augmented reality**, in which pure reality is mixed or blended with virtual reality. VR applications include flight simulators used to train pilots, architectural or geological visualisations, and electronic entertainment, especially video games. In recent years, interest in VR has skyrocketed: Oculus Rift and other devices have opened new doors for games, films, and design processes by combining simple devices with tremendous computing power.

BIOGRAPHIES

5VOLTCORE

In 2003, Emanuel Andel and Christian Gützer, then classmates at the University of Applied Arts in Vienna, created the artist collective 5VOLTCORE. The since disbanded group exhibited various media installations and performances, examining the behaviour of technology, its redundancies, and the issue of trust. In 2004, they received recognition for *Shockbot Corejulio,* a self-hacking and subsequently self-destructive computer/robot arm. In 2007, Andel and Gützer created the *knife.hand.chop.bot,* which was nominated for the transmediale award in 2008 and can be seen in *Hello, Robot.*

AKA Intelligence

Founded in 2013 in Wilmington, Delaware, USA, the company develops artificial intelligence (AI) in order to improve the communication between objects as well as between objects and humans. In the process, AKA merges AI and Big Data and designs communication tools such as language, writing, facial expressions, and gestures, all of which provide training and educational robots with the skills they need in order to do their job.

ABB

The ABB Group is pioneering technology leader in electrification products, robotics and motion, industrial automation and power grids. Headquartered in Zurich, Switzerland, it serves customers in utilities, industry, and transport and infrastructure globally. The present company was formed in 1988 from the merger of two pioneering electrical engineering companies: ASEA of Sweden and BBC of Switzerland. In 2015, the group employed 135,800 people worldwide and generated just under 35.5 million USD in revenues. With more than 70 million connected devices and over 70,000 industrial control systems installed globally, ABB is at the forefront of the energy and fourth industrial revolutions. Its solutions cover all points of electricity consumption, and extend to industrial robots, electric motors and drives, navigation systems for ships, turbo chargers, and power grids. ABB has the world's broadest robotics portfolio, and an installed base of over 300,000 robots worldwide.

Aktion Mensch e.V. and Jan Hinrik Drevs

Inclusion is the main focus of the social organisation Aktion Mensch e.V. (founded in 1964, located in Mainz, Germany), which is financed through lottery earnings. The campaign film *Neue Nähe* was made by the documentary, feature, and advertising filmmaker Jan Hinrik Drevs, who has worked mainly as a freelance author and director for television and film since 1998. For *Neue Nähe,* Drevs worked with a team of interested amateurs – none of whom are professional actors.

Woody Allen

(b. 1935 as Allan Stewart Konigsberg in New York, USA) is an American comedian, film director, author, actor, and jazz musician. After beginning his career as a columnist and stand-up comedian, from the mid-1960s onwards, Allen went on to write screenplays for and direct over fifty films, all while countless writing stories and plays. He has received four Oscars and twenty-four Oscar nominations for his films, most of which are comedies marked by his characteristically dark undercurrents and satirical notes. With one film released per year for nearly the last forty years, Allen ranks among the most prolific directors of all time.

Amazon

The publically-owned American online mail-order company and service provider, founded in 1995 by Jeff Bezos as an online bookshop, is the market leader in many sectors today. It has expanded its business area to nearly every product group and sells electronic devices such as e-readers and speech recognition systems under its own brand. Moreover, Amazon runs platforms for music and video downloads and online payment services, etc. The global enterprise often comes under criticism for issues such as market abuse, tax avoidance, and the exploitation of its employees.

Archigram

Avant-garde group of architects formed in London in 1961 and disbanded in 1974, published a magazine of the same name. They saw themselves as consumer-oriented Neo-Futurists, and developed fictional projects pointing the way forward to a future, highly technological reality. Archigram's core members were Ron Herron (b. 1930 in London, d. 1994 in Woodford, London), Warren Chalk, David Greene, Dennis Crompton, Peter Cook, and Michael Webb.

Isaac Asimov

(b. 1920 in Petrovichi, Russia, d. 1992 in New York, USA) was an American scholar, a professor of Biochemistry at Boston University, and a prolific author. He is mostly known for his science fiction and popular science writings, but he published in many genres. His oeuvre includes seminal sci-fi works like *I, Robot* and the *Foundation* series. Considered one of the best science fiction writers during his lifetime, Asimov won several Hugo and Nebula Awards.

asmbld.

Robotics start-up headquartered in Brooklyn, New York, founded in 2014 by Petr and Ted Novikov. The company specialises in solutions and new approaches to automation in the construction trade using new technologies and robotics and works together with construction and real estate companies. The company believes that robotic technology has a huge potential for reducing costs through speed and improved safety not only in the manufacturing sector, but also in the construction industry.

automato.farm

The Shanghai-based studio for design research was founded in 2015 by Simone Rebaudengo, Matthieu Cherubini, and Saurabh Datta. With a professional background in computer science, product design, and electrical engineering, automato.farm develops prototypes and installations, undertakes design research and investigates, partially in a speculative way, the impact of new technologies on our daily lives.

Bandai

The leading toy manufacturer in Japan and the fourth largest toy maker in the world. Established in 1950 in Tokyo, Japan, the company also produces video games, anime, tokusatsu programs, and plastic model kits. In 2005, Bandai merged with Namco, a video game, arcades and amusement parks specialist, becoming part of the Bandai Namco Group. After the restructuring of the group in 2006, Bandai now operates the Toys and Hobby Strategic Business Unit and owns the main toy licenses in Japan, including those for Ultraman, the Power Rangers series, and Gundam.

Bruno Bianchi

(b. 1955 in Tours, France, d. 2011 in Paris, France) was a French cartoonist and animator. In 1977, he joined the animation department of DiC Entertainment, where he worked as producer, artist, animator, television and film director, and writer. He is mostly known for his work on *Inspector Gadget,* an iconic television cartoon series he co-created with Andy Heyward (b. 1949 in New York, USA), DiC's former Chairman and CEO and DiC's founder Jean Chalopin (b. 1950 in Tours) in the 1980s. Bianchi was the main character designer and supervising director of the series, which ran for two seasons from 1983 to 1986. The cartoon continued to air in reruns, and grew into a comic book series and a film by Disney in 1999.

Philip Beesley

(b. 1956 in Westcliff-on-Sea, Great Britain) is a Canadian architect, visual artist, and Professor in Architecture at the University of Waterloo and Professor of Digital Design and Architecture & Urbanism at the European Graduate School. His work focuses on embedding interactivity into spatial environments. Since 2001, he has been working in the emerging world of digital fabrication and smart materials to create spaces where users can interact with environments that react and evolve. He is best known for his robotic rainforests, jungle-like environments filled with robotic elements designed to interact with one another. His work has been displayed in numerous museums and biennials such as the 2011 Venice Architecture Biennale.

Wafaa Bilal

(b. 1966 in Najaf, Iraq) is an artist and professor. Bilal is known for his public performances and technology-based, interactive artworks. With a highly political perspective, the artist aims to initiate discussions about violent international conflicts. His work is shaped by his experience of fleeing his home country during the first Gulf War. After living in refugee camps for two years, he was granted asylum in the USA. He currently teaches at New York University's Tisch School of the Arts.

Studio Bitonti

Founded in 2007 by the designer Francis Bitonti (b. 1983, New York State, USA), the New York-based Studio Bitonti is committed to the application of digital design and computer driven manufacturing technologies. It uses cutting-edge techniques to create intricate objects without human manual labour, thereby transforming the logic of traditional mass production. The studio first gained recognition for a 3D printed dress designed for Dita von Teese.

Otto Bock Healthcare

In 1919, the Thuringia-born prosthetist and entrepreneur Otto Bock founded the Orthopädische Industrie GmbH, allowing him to cater to the rapidly rising demand for prostheses and orthopaedic products after the end of the First World War. The company experimented early on with innovative materials, such as aluminium (at the start of the 1930s), and it still sets the bar in the global world of prosthetics by incorporating modern technology and innovative design into its product development.

Bureau d'études

Founded in 1998 by Léonore Bonaccini and Xavier Fourt. In recent years they have produced cartographies of current political, social, and economic relationships. These visual analyses of transnational capitalism are based on extensive research and are presented in the form of large-format wall charts. They make the invisible visible and relate the apparently unrelated in an illuminating fashion.

Julius Ingemann Breitenstein

Born in Copenhagen, Denmark, in 1995 and raised in London, Julius Breitenstein studied Product Design at London's Central Saint Martins, graduating with a BA in 2016. The designer describes his work as "harmonizing machine learning with user experience and product design." While Breitenstein focuses his attention on efficiency within the process of designing (drawing influence from "desire paths"), he also challenges the notion of algorithms as the be-all and end-all of creating, and advocates for designers to further develop machine-based results. Breitenstein lives and works in Copenhagen and London.

Björk Guðmundsdóttir

(b. 1965 in Reykjavík, Iceland) is an Icelandic singer, composer, music producer, and actress. Her work reflects her interests in a number of musical genres – from pop and electronic music to trip-hop and alternative rock, jazz, and folk – as well as her delight in costumes, transformation, and performance. The inimitable artist has sold over twenty million albums worldwide.

Sander Burger

(b. 1975, Ivory Coast) is a Dutch director and documentary filmmaker. He studied directing at the Netherlands Film Academy and has since produced a number of short films, short documentaries, and works for television. His film *Hunting & Zn.* was nominated twice for the prestigious Dutch Golden Calf award in 2010; his film *Ik ben Alice* (English: *Alice Cares*) premiered at the International Film Festival Rotterdam in 2015.

Edward Burtynsky

(b. 1955 in St. Catharines, Canada) is an artist most well known for his large-format photographs depicting landscapes overcome by industrial growth, a motif he has dedicated himself to since the 1980s. As a child Burtynsky developed black-and-white films in an in-home darkroom together with his father. After studying graphic arts and photography in Ontario, in 1985 he opened the Toronto Image Works, a photography centre and lab. Today he is an internationally renowned photographer and is the recipient of numerous honorary doctorates as well as the 2005 TED prize.

James Cameron

(b. 1954 in Kapuskasing, Ontario, Canada) is a director, producer, and screenwriter who specialises in action and science-fiction films. He became famous thanks to blockbusters such as *Alien, Terminator,* and more recently *Avatar,* which became the highest-grossing film of all time.

Zackary Canepari and Drea Cooper

Photographer Zackary Canepari (b. 1979, Boston, Massachusetts, USA) and film-maker Drea Cooper (b. 1977, Honolulu, Hawaii, USA) teamed up in 2009 and have been working together ever since. Appearing on *Filmmaker Magazine*'s list of "Top 25 New Filmmakers to Watch", the pair mostly shoots series of discrete short documentaries. Along with *The Family Dog,* the *ROBOTICA* series they produced for *The New York Times* in 2015 also includes episodes on sex androids and the bomb disposal robots deployed by the US military.

Gonçalo F. Cardoso and Ruben Pater

Gonçalo F. Cardoso (b. 1979 in Lisbon, Portugal) is a London-based disc jockey and sound artist. He initiated Discrepant.net as a platform for releasing music by like-minded artists who, like himself, aim to "deconstruct, distort and re-assemble the lore of (un)popular music." Ruben Pater (b. 1977 in Gouda, The Netherlands) is a designer and researcher from Amsterdam with a background in graphic design and liquid journalism. His design work under the name *Untold Stories* is highly political.

Dan Chen

(b. 1982, Tainan, Taiwan) is a designer and engineer. He has been working at the intersection of robotics, communication design, interaction design and product design for years, partially thanks to his interdisciplinary education from such prestigious universities as RISD (The Rhode Island School of Design) and MIT (The Massachusetts Institute of Technology). Today, he researches and evaluates – in a critical way – contemporary forms of communication, as well as human experience in light of newer, more networked and more roboticised communication environments. Chen works for IDEO in New York City.

Citroën / Euro RSCG

Citroën, founded in 1919, is a leading French car manufacturer. For the launch of their C4 compact car, the company brought together a multidisciplinary group of collaborators. The London-based advertising agency Euro RSCG (rebranded in 2012 to Havas Worldwide London) was hired to lead a diverse team including director and CG artist Neill Blomkamp (known for the popular sci-fi film *District 9*) as well as choreographer to the stars, Marty Kudelka.

Arthur C. Clarke Center for Human Imagination

The Arthur C. Clarke Center for Human Imagination at the University of California, San Diego was initiated by a grant from the Arthur C. Clarke Foundation in 2012. The Clarke Center is under the direction of American media artist and educator Sheldon Brown. In the summer of 2016, Brown worked together with authors (and Casa Jasmina co-founders) Bruce Sterling and Jasmina Tešanović during their residency at the Clarke Center to create the media installation *My Elegant Robot Freedom* specifically for the exhibition *Hello, Robot*. Additional collaborators on the project were Amanda Bergman, Lyndsay Bloom, Wes Hawkins, Erik Hill, Jon Paden, Pepe Rojo, Rosanna Viirre, Aleksander Viirre, Ash Smith, and Nathan Wade.

Carnegie Mellon University

Founded in 1900, Carnegie Mellon University (CMU) in Pittsburgh, Pennsylvania is one of the most renowned research institutions in the USA. It is a world leader specifically in the fields of computer science and robotics. The National Robotics Engineering Center at the Robotics Institute developed, amongst others, the military robot *Dragon Runner* in cooperation with the US Marine Corps Warfighting Lab. The research project was led by Prof. Hagen Schempf, the former director of the Hazardous Environments Robotics Lab at CMU and a member of the original team of roboticists who discovered the *Titanic* and the *Bismarck*. Prototypes of *Dragon Runner* were first developed by Schempf's start up, Automatika, Inc.; today, the robot is distributed by QinetiQ North America.

Karel Čapek

(b. 1890 in Klein Schwadowitz, Austria-Hungary [now Malé Svatoňovice, Czech Republic], d. 1938 in Prague, Czechoslovakia) is one of the most important Czech authors of the past century. Many of his works, which are a testimony to his keen observation skills as well as his satirical humour, focus on the ethical aspects of societal developments such as mass production or weapons of mass destruction. During the 1930s he wrote against fascism and nationalism. Today Čapek is primarily known around the world as a pioneer in the genre of science fiction; the term "robot" first appeared in his technological dystopian drama *R.U.R. (Rossum's Universal Robots).*

Arthur C. Clarke

Arthur C. Clarke (b. 1917 in Minehead, Somerset, Great Britain; d. 2008 in Colombo, Sri Lanka) began writing science fiction in 1937. Thanks to his particular interest in the possibilities of space exploration, he is seen as one of the genre's technical visionaries. Clarke also studied mathematics and physics, working as a scientist throughout his lifetime. Outside of the sci-fi community, his name is primarily associated with the success of *2001: A Space Odyssey*. Inspired by one of the author's short stories, the screenplay was a collaborative effort between Clarke and the director Stanley Kubrick.

Clouds AO and SEArch

Clouds Architecture Office was founded in New York in 2010 by Ostap Rudakevych, Masayuki Sono, and Yuko Sono. The design for the Mars Ice House was developed in collaboration with Christina Ciardullo, Kelsey Lents, Jeffrey Montes, Michael Morris, and Melodie Yashar from Space Exploration Architecture, a collaborative effort between architects and designers focussing on space exploration and praxis.

Manu Cornet

(b. 1981, Paris, France) is a programmer, cartoonist, author, and musician based in San Francisco. Aside from working full-time as a software engineer at Google, Cornet is also known for his book *The Crab and the Lamb* as well as for his satiric cartoons and graphics. The latter often revolve around people's dealings with new technology and have been published in *The New York Times* and *Der Spiegel,* amongst others.

Jan De Coster

(b. 1973 in Brakel, Germany) completed his education in the fields of physics and engineering. After shifting gears to work on online advertising campaigns, he discovered his talent for story-telling and creating lively, empathetic characters. In 2008, De Coster opened his studio Slightly Overdone, where he develops robots for installations, exhibitions, and workshops. He has toured the globe to teach on the intersection of electronics and art, always with his personable robots in tow.

Douglas Coupland

(b. 1961, Rheinmünster-Söllingen, Germany) is a Canadian author and visual artist based in Vancouver. His first novel, *Generation X* (1991), became an international bestseller and introduced terms like "McJob" and "Generation X" to the general public's vocabulary. To date, Coupland has published 15 novels and various collections of short stories. His writing contains postmodern descriptions of states of affairs, revealing aspects of pop culture, life in Web 2.0, or sexuality in a sociologically sensitive way. Coupland's bold visual artworks often make references to these same themes.

Chris Cunningham

(b. 1970 in Reading, Great Britain) is a British director with a particular focus on music videos, commercials, and video art. He made a name for himself with his often surrealist music videos for artists such as Portishead, Madonna, Björk, and Aphex Twin. Cunningham also creates video installations and is an expert on special effects for large-scale film productions such as *Alien 3* or *Judge Dredd.*

CurVoxels

The MArch Architectural Design programme at the Bartlett School of Architecture at UCL is organised in research clusters that approach architecture in speculative ways. CurVoxels was formed within the framework of Research Cluster 4 by the students Hyunchul Kwon, Amreen Kaleel, and Xiaolin Li. Each year, students are presented with the challenge of creating a cantilever chair based on rapid prototyping. For their chair, CurVoxels received the Gold Track Award for the best project of the year.

Daft Punk

The band, founded in 1993, is made up of French musicians Guy-Manuel de Homem-Christo (Guy-Man, b. 1973 in Paris, France) and Thomas Bangalter (b. 1975 in Paris, France). The duo creates music that blends house, electronic, techno, rock, funk, and synthpop and is known for their sleek and enigmatic "robot" stage personas. In 2003, Daft Punk teamed up with the Japanese animation studio Toei to create the visualisation for their 2001 album *Discovery.* The film was directed by Kazuhisa Takenouchi with artwork by renowned anime artist Leiji Matsumoto. From this work the anime film *Interstella 5555 – The Story of the Secret Star System* evolved.

Dangerous Things LLC

The young company, founded by Amal Graafstra in 2013 and headquartered in Bellingham, Washington, USA, develops and markets miniaturised data chips for implantation under the skin as well as related equipment such as sterile scalpels and "injection kits". Dangerous Things propagates the current and future possibilities of biohacking and of the cyborg, a hybrid of living organism and machine. Among his other appearances, Graafstra was a guest at the CeBIT 2016.

Dunne & Raby

For years, designer Anthony Dunne (b. 1964 in London, Great Britain) and architect Fiona Raby (b. 1963 in Singapore) have been at the forefront of a conceptual design movement for which ground-breaking ideas and debates are more important than functionality. Dunne and Raby are particularly interested in the design potential and everyday impact of new technologies such as robotics and bio- or nanotechnology. Before moving to New York to take on their new roles as professors of Design and Emerging Technology at Parsons School of Design at the New School in 2016, Raby was professor for Industrial Design at the University of Applied Arts in Vienna and Dunne was the Head of the Design Interactions programme at RCA in London.

École cantonale d'art de Lausanne (ÉCAL)

ÉCAL was founded in Lausanne, Switzerland in 1821 and ranks among Europe's most renowned art academies. Now led by Alexis Georgacopoulos, it offers six bachelor's and four master's programmes in the fields of visual arts, film, graphic design, industrial design, photography, and media and interaction design. The academy, which maintains close ties with industry and high-ranking cultural institutions, has produced a number of well-known designers such as BIG-GAME and Adrien Rovero as well as artists including Cyprien Gaillard.

Ekso Bionics

Founded in 2005, the California-based company produces medical devices that enhance patients' mobility using robotics. In 2009 Ekso Bionics launched the Ekso GT™ robotic exoskeleton, a product ranked by *Time Magazine* as one of the "50 Best Innovations of 2010": the first licensed robot skeleton that helps paralysed patients to get up and walk by themselves.

Tal Erez

The designer, curator and researcher (b. 1981 in Ramat Gan, Israel) studied industrial design at the Holon Institute of Technology and received his MA in Conceptual Design from the Design Academy Eindhoven. His design office "Tal Erez: Design Related" has developed various international exhibitions including the Israeli Pavilion for the 13th Venice Architectural Biennale. Erez teaches at the Bezalel Academy of Art and Design's MA Program in Jerusalem and heads the city's Design Week. Erez lives and works in Israel and the Netherlands.

Federal Art Project

The Federal Art Project (1935–43) was an American New Deal subsidy programme for the visual arts during the administration of President Franklin D. Roosevelt. Its primary goal was to provide work to artists who had been unemployed since 1929 during the Great Depression. Over 100 Community Art Centres were established throughout the country, offering spaces where they could work, exhibit, and teach. The programme provided work for over 10,000 artists.

Festo

Based in Esslingen am Neckar, Germany, Festo is a leading international company in the fields of factory and process automation. A key focus of its extensive research and development activities is bionic learning, the mimicking and modelling of natural processes and their utilisation in new technological solutions. The company was founded in 1925 and is today a global player with over 250 branches and service offerings in 176 countries.

Flower Robotics

The company, founded in 2001 by Tatsuya Matsui (b. 1969 in Tokyo, Japan), develops robots for domestic life as well as for the commercial sector (e.g. robotic mannequins). The aim and philosophy of Flower Robotics is to create robots which are "natural" and integral, yet subtle, accompaniments to our everyday life. Through a focus on aesthetics, the emotional content of communication and the latter's added value for the user, the company is designing our gradually more robot-filled daily life.

Flying Lotus and Beeple

The DJ, musician, and producer Flying Lotus – born in 1983 as Steven Ellison in Los Angeles – made a name for himself with his Hip-Hop and experimental electronic music. Mike Winckelmann works under the name Beeple. According to his website, he produces "a variety of art crap across a variety of media" often focusing on humans and machines: short films, VJ loops, or music videos, such as his video for Flying Lotus' *Kill Your Co-Workers*.

Vincent Fournier

(b. 1970 in Ouagadougou, Burkina Faso) is a Paris-based photographer. After studying sociology and art, he graduated from the École Nationale de la Photographie in 1997. He originally worked in the advertising industry before focusing on his artistic work some years ago. He is known for his photo series dealing with modern science and technology, such as *Man Machine, Natural History,* and *Space Project.*

Alex Garland

(b. 1970 in London, Great Britain) is a British author, screenwriter, and director with a BA in art history. His debut novel, *The Beach,* quickly became a bestseller and was later filmed, as was his second novel *Manila.* In addition to screenplays for science-fiction and literary films, Garland made his debut as a director in 2015 with the film *Ex Machina.*

Yves Gellie

(b. 1953 in Bordeaux, France) is a French photographer. After studying medicine and working in the field of tropical medicine, he began his photojournalistic career in 1981 with a feature on Colombian cocaine production before moving on to cover the story of Somali refugees. His works, a mixture of photojournalism and visual art, are exhibited internationally.

Greenpeace

The non-profit organisation, founded in 1971 by Canadian peace activists, has approximately three million supporting members worldwide and circa 2,400 employees. Greenpeace began by planning actions against the United States' planned atomic bomb tests in Alaska and the Pacific before expanding its operations to cover increasingly international issues such as global warming, biodiversity, species protection, and green genetic engineering. Greenpeace is the world's largest environmental organisation.

Matt Groening and David X. Cohen

Matt Groening (b. 1954 in Portland, Oregon, USA) is a cartoonist who joined forces with scriptwriter and film producer David X. Cohen (David Samuel Cohen, b. 1966 in New York, USA) to collaborate on *Futurama* in 1997. Cohen also wrote for several episodes of *The Simpsons,* an animated series created and drawn by Groening. Both series enjoy immense popularity worldwide and have won numerous awards for writing and animation.

Kevin Grennan

(b. 1984 in Galway, Ireland) is a user experience designer who lives and works in San Francisco. Grennan studied communication design before earning his master's degree in Design Interactions at London's Royal College of Art. His work ranges from the development of mobile apps and interface design to experimental works in the fields of video and graphic design. Grennan became a designer for Google Maps in 2011.

Gramazio Kohler Research

Fabio Gramazio (b. 1970 in Walkringen, Switzerland) and Matthias Kohler (b. 1968 in Uster, Switzerland) founded the architectural studio Gramazio Kohler Architects in Zurich in 2000. In their work they complement traditional design, construction, and building methods with the possibilities provided by computers and digital fabrication techniques. The sensible qualities of this design style are manifested in the new expression of a "digital materiality". As of 2005, Gramazio and Kohler have been chairs of the Department of Architecture and Digital Fabrication in the Institute of Technology in Architecture at the ETH Zurich, also known as Gramazio Kohler Research.

Vicente Guallart and IAAC

The architect Vicente Guallart (b. 1963 in Valencia, Spain) opened his Barcelona studio in 1993 and later held the position of the city's Chief Architect from 2011 until 2015. In 2011, he founded the Institute for Advanced Architecture of Catalonia (IaaC), which organises a multidisciplinary programme of events dedicated to promoting sustainability as well as innovative construction methods such as digital fabrication. The IaaC and Guallart Architects teamed up with MIT's Center for Bits and Atoms and the design company Bestiario to create the installation *Hyperhabitat: Reprogramming the World* for the 2008 Venice Architecture Biennale.

Hanna-Barbera Productions

American animation studio founded in 1957 by the former Metro-Goldwyn-Mayer animation experts William Hanna and Joseph Barbera, the creators of *Tom and Jerry* as well as other animated hits. Their popular series such as *The Jetsons, The Flintstones,* or *Scooby-Doo* allowed Hanna-Barbera to dominate the animation segment of American television for over three decades during the second half of the twentieth century.

John Heartfield

The artist John Heartfield was born Helmut Herzfeld in Berlin in 1891 (d. 1968 in the former GDR). After founding the political publishing house Malik with his brother in 1917, Heartfield put his layouting skills to work for the Berlin Dada group, where he began a prolific career in political photomontage. With his eye for detail, sharp wit, and communist leanings, Heartfield redefined the medium and offered a fresh perspective to the otherwise disillusioned working classes.

Guy Hoffman

(b. 1973 in Jerusalem, Israel) is an assistant professor at Cornell University in Ithaca, New York, USA, where he heads the field of work for Human-Robot Collaboration and Companionship. He researches algorithms, schemes of interaction and designs that enable close exchanges between humans and robots both in the workplace and at home. Amongst others, he developed the first theatre performance featuring humans and robots as well as the first improvising human-robot jazz duo. Hoffman holds higher-education degrees in the fields of both human-robot interaction and computer science, but also studied animation at Parsons School of Design in New York.

Susanna Hertrich

(b. 1973 in Paris, France) is a German artist working at the intersection of art and technology. She invents wondrous tools and wearables, which satirically comment on our highly technological society in a subtle way. Her body prostheses are a critique of society and a representation of an alternative future all at once. Hertrich lives in Berlin and Basel.

Holland Haptics

Holland Haptics was founded in 2012 by Frederic Petrignani, Michel Dufrasnes, and Arun Sadhashivan as a consumer electronics company invested in allowing people to experience the sense of touch via the Internet during online conversations. Holland Haptics created *Frebble,* a computer accessory which enables a virtual hand-holding experience.

Höweler + Yoon

Founded in 2004 by Eric Höweler (b. 1972 in Cali, Colombia) and Meejin Yoon (b. 1972 in Seoul, Korea), Höweler + Yoon is an internationally renowned architecture and design studio based in Boston. Its multidisciplinary activities range between architecture, art, and landscape planning while integrating various media and working at the interface of the conceptual and the tangible. Work by Höweler + Yoon has won numerous awards and has been exhibited in museums including the MoMA and the Guggenheim in New York.

Ted Hunt, Luke Sturgeon and Hiroki Yokoyama

Ted Hunt (b. 1976, Wales, Great Britain), Luke Sturgeon (b. 1984 in Stamford, Great Britain) and Hiroki Yokoyama (b. 1977 in Gunma, Japan) studied Design Interactions together at the Royal College of Art in London, where they explored the subject of acceptance of and interaction with drones in our daily lives. While Hunt and Sturgeon still operate in the field of interaction design today, Yokoyama now specialises in filmmaking.

Zan-Lun Huang

(b. 1979, Yilan County, Taiwan) is an artist whose practice responds to the hybridisation of robotic and organic lives, and who aims to discuss the implications of technological development on human nature and desire. Huang studied fine arts at Taipei National University of the Arts. His works have been exhibited extensively around the world, with solo exhibitions at the Taipei Artist Village and Kuandu Museum of Fine Arts, Taipei.

ICD/ITKE Stuttgart

Since its founding in 2008 by the architect Achim Menges (b. 1975 in Mannheim, Germany), the Institute for Computational Design (ICD) at the University of Stuttgart has quickly become one of the leading graduate programmes for machine computational design and robotic fabrication worldwide. The ICD cooperates closely on an annual research pavilion with the Institute of Building Structures and Structural Design (ITKE) at the University of Stuttgart, led by structural engineer Jan Knippers (b. 1962 in Düsseldorf, Germany). Both institutions collaborate extensively with researchers from the biological sciences and have a strong focus on materials science.

Keiji Inafune and Capcom

The video game developer Capcom (founded 1979 as I. R. M. Corporation in Osaka, Japan) specialised in arcade games prior to the release of the home console game *Mega Man* in 1987. For this launch, the company enlisted senior Capcom employee Akira Kitamura to create the pixel art Sprite for Mega Man. The then unknown artist Keiji Inafune (b. 1965 in Kishiwada, Japan), now a prominent video game producer and illustrator, founder of Comcept, was hired to create the rest of the artwork. Since the launch of the original game, countless sequels and spin-offs have been released amid growing international popularity.

Interactive Architecture Lab

The Interactive Architecture Lab at University College London's Bartlett School of Architecture is a multidisciplinary research laboratory which deals with robotic, responsive environments, wearable computing, the Internet of Things and similar topics. The lab is closely connected to a 12-month master's programme covering these issues, and it communicates and exchanges ideas with other research & development institutions as well as the industry on a regular basis.

Alfredo Jaar

The artist, architect, and filmmaker Alfredo Jaar (b. 1956 in Santiago de Chile, Chile) moved to New York in 1982 and began creating powerful, socially motivated work that surpassed cultural and national boundaries; he devoted himself to such daunting topics as the Serra Pelada goldmine in Brazil in *Gold in the Morning* (1986) or the Rwandan genocide in the longstanding *Rwanda Project* (1994–2000). Along with his prolific body of film and photography work, Jaar stages public interventions which, famously, make participants out of his spectators, who are given no option but to question the status quo together with the artist.

Spike Jonze

The 2013 film *Her* was the screenwriting debut of Spike Jonze (b. 1969 in Rockville, Maryland, USA, as Adam Spiegel), who also directed and produced the film. He had already achieved directing success with hit films such as *Being John Malkovich* (1999) and music videos for well-known bands and performers such as Daft Punk, the Beastie Boys, and Björk. He came up with the idea for *Her* in 2000 after reading an article about a website offering live chats with artificial intelligences.

Floris Kaayk

An artist and film director, Floris Kaayk (b. 1982 in Tiel, the Netherlands) studied art in Breda and Amsterdam and has become well known for his fictional documentaries and mockumentaries, all of which have quickly gone viral on the Internet. In 2016 he won the Golden Calf award at the Dutch Film Festival for his project *Oscar: The Modular Body*.

Friedrich Jakob Kiesler

(b. 1890 as Friedrich Jakob Kiesler in Czernowitz, Austria-Hungary [now Chernivtsi, Ukraine], d. 1965 in New York, USA) was an Austrian-American architect, artist, designer, set designer, and exhibition organiser. His utopian, usually trans-disciplinary works and concepts influenced the avant-garde of his era and continue to provide an important impulse to this day. Examples include his pioneering concept for a stage open on all sides, his theory of correalism, with which he relates people, artworks, and the environment to one another, as well as his concept for the Endless House.

iRobot

American, market-listed producer of robots, which was founded in 1990 by the three former MIT employees Rodney Brooks (b. 1954 in Adelaide, Australia), Colin Angle, and Helen Greiner. After having long-standing success in the fields of military and security robots, the company now focuses mainly on domestic robots and technologies for home automation. According to their own statistics, iRobot sold 14 million domestic robots in the year 2016.

Kiiroo

Founded in 2013 and headquartered in Amsterdam, Netherlands, Kiiroo is a company specialising in virtual-reality sex. It began by developing interactive sex toys for couples in long-distance relationships. They soon discovered, however, that their products were compatible with virtual-reality films, adding physical sensation to the experience of pornographic films by making the movements of the toys correspond to the images on screen. According to the manufacturer, young people in their twenties and thirties are the main target audience for this technology.

Elizabeth King

(b. 1950 in Ann Arbor, Michigan, USA) is an American sculptor. Her works, which combine moving, figurative sculptures with stop-motion animation, blur the boundaries between real and virtual objects. They reflect the artist's interest in early automatons, the history of puppets, and literary stories in which artificial figures come to life. King is a lecturer in the Department of Sculpture and Extended Media at Virginia Commonwealth University. For the work "What Happened" she teamed up with artist and filmmaker Richard Kizu-Blair.

Kram / Weisshaar

Design studio founded in 2002 by Reed Kram (b. 1971, Columbus, Ohio, USA) and Clemens Weisshaar (b. 1977, Munich, Germany) with headquarters in Munich and Stockholm, Sweden. Their projects range from software development and process design to product design and architecture, focusing on the slowly dissolving boundary between the realms of the digital and the physical.

KUKA AG

KUKA Robotics Corporation is a subsidiary of the KUKA AG, which was founded in 1898 by Johann Josef Keller and Jakob Knappich in Augsburg, Germany. In 1973, the company introduced FAMULUS, the world's first industrial robot with six electromechanically driven axes. KUKA has since become one of the leading industrial robot and automation manufacturers in the world. To this day, the company is based in Augsburg, which is midway between the respective headquarters of automobile industry giants – and KUKA customers – BMW (in Munich) and Mercedes-Benz (in Stuttgart).

Kraftwerk

The band, a pioneering force in the electro-pop scene, was founded in Dusseldorf by Ralf Hütter (b. 1946 in Krefeld, Germany) and Florian Schneider (b. 1947 in Öhningen, Germany) in 1970. After producing three acoustic albums Hütter and Schneider switched to purely electronic music in 1973. Their next record, *Autobahn,* is considered the world's first electro-pop album. The album *Mensch-Maschine,* released in 1978 after the band had grown to four members, and the 1981 album *Computer-welt* are likewise considered important trailblazers in the development of electronic music.

Stanley Kubrick

The films of the American director, screenwriter, and producer Stanley Kubrick (b. 1928 in New York, USA, d. 1999 in St. Michael, Great Britain) focus on his protagonists' battles with the dark side of human nature, with human impulses, dreams, and reality, and are remarkable for their frequently allegorical imagery. Even today, his films such as *2001: A Space Odyssey,* *Clockwork Orange,* or *Eyes Wide Shut* are the topic of lively discussion and are regarded as cult films by growing audiences.

Joris Laarman

(b. 1979 in Borculo, the Netherlands) is a Dutch designer, artist, and entrepreneur known for his experimental designs as well as his inspiration by and use of coming new technologies. Laarman, a graduate of the Design Academy Eindhoven, achieved his international breakthrough with his *Heatwave Radiator,* which was manufactured by Droog. Together with his partner Anita Star, Laarman has directed the Joris Laarman Lab in Amsterdam since 2004.

Fritz Lang

(b. 1890 in Vienna, Austria, d. 1976 in Beverly Hills, California, USA) was an Austrian-German director, screenwriter, and producer. He played a major role in shaping cinema in two countries: Germany in the 1920s and early 1930s (with films such as *Metropolis* and *M*), and the USA from the late 1930s to the 1950s (with films such as *Scarlet Street*). Lang was already an established director in 1934 when he emigrated to the United States and embarked on a 20-year career in Hollywood. The Expressionism of his films is considered integral to the evolution of American genre cinema, particularly film noir.

Glen A. Larson

(b. 1937 in Los Angeles, California, d. 2014 in Santa Monica, California, USA) made a name for himself as a screenwriter and producer of some of the most well-known American television series ever, including *Quincy, M.E.* (1976), *Battlestar Galactica* (1978), *Magnum, P.I.* (1980), and *Knight Rider* (1982). Part of these shows' success was the fact that, thanks to straightforward plots, endearing main characters, a great deal of humour, and little violence, they were highly watchable by the entire family.

George Lucas

(b. 1944, Modesto, California, USA) is an American director, screenwriter, producer, and businessman. He created the vastly successful six-part *Star Wars* film series as well as the *Indiana Jones* movies. Having finished his studies in film at the University of Southern California, Lucas is known as one of the pioneers of the digital cinema camera. Today he is one of the most successful members of the film industry, owning an animation studio as well as the film and television production company Lucasfilm. With assets estimated at three billion euros, Lucas is one of the world's richest people.

John Lasseter

(b. 1957, Los Angeles, California, USA) is the chief creative officer of Pixar Animation Studios. He was the second student ever to enrol in the famous character animation programme created by veteran Disney animators at The California Institute of the Arts (CalArts), and the first animator at Pixar Studios. Lasseter began his career as an animator with The Walt Disney Company. He later joined Lucasfilm, where he worked on the then radical new medium of computer-generated imagery (CGI) for animation. The Graphics Group of the Computer Division of Lucasfilm was sold to Steve Jobs and became Pixar in 1986. Lasseter wrote, directed, and animated Pixar's first short films, including *Luxo Jr.*

Leka SAS

Paris-based start-up founded in 2014 by Ladislas de Toldi and Marine Couteau focusing on the development and marketing of a therapeutic toy of the same name. Their multi-sensor, robotic ball named *Leka* was created for children with special needs, such as autism, in order to encourage their interests and motivations both at home and in therapy.

Greg Lynn

(b. 1964 in North Olmsted, Ohio, USA) is an American architect and science fiction author. A graduate of Princeton University, he is famous for his biomorphic architecture created with the help of computer-aided design. He developed this approach in 1993 in his book *Folding in Architecture.* He is the director of the Greg Lynn Form architecture studio in Venice, California, and lectures at several institutions, including the University of Applied Arts in Vienna and the UCLA School of Arts and Architecture. In 2008, he was awarded the Golden Lion at the Venice Architecture Biennale.

Aki Maita

An employee with the Japanese toy manufacturer Bandai, Aki Maita (b. 1967) noted that girls like looking after animals – living ones and toy ones alike – and that they particularly love things that fit in their trouser pocket. At the beginning of 1996, she submitted the idea for Tamagotchi to her superiors. By 2010, Bandai had already sold over 76 million of the "computer eggs" on the global market.

Keiichi Matsuda

(b. 1984 in Hong Kong) is a British-Japanese designer and filmmaker with a studio in London. In his often internationally exhibited and projected works, he investigates the effects of new, upcoming technologies on the perception and living environment of humans. Interested in the dissolution between physicality and virtuality, Matsuda works in the fields of video, architecture, and interactive media.

Shawn Maximo

(b. 1975 in Toronto, Canada) is a New York City-based artist and architect. Native to the gateways of sculpture, digital media, and design, Maximo creates experimental settings and virtual renderings in collaboration with the collective Yemenwed. He also designs shop-windows for well-established brands. Maximo's works are characterised by the interaction and confrontation of architectural space with seemingly unfamiliar functions, which initially have a rather disturbing effect on people but unfold their visionary power at second glance.

Moth Collective

Oft-awarded animation studio, which was founded by the former Royal College of Art students Daniel Chester, Dave Prosser, and Marie-Margaux Tsakiri-Scanatovits in 2010 in London. Especially vaunted for their subtle, hand-drawn, thought-stimulating animations, Moth Collective works with clients like *The New York Times,* Kiehl's, *The Guardian* and others.

NASA

The National Aeronautics and Space Administration (NASA) is an independent U.S. government agency responsible for the space programme. As the global leader in space exploration, NASA mounted the Mars Science Laboratory mission (MSL, launched in 2011), which landed the Curiosity rover on the red planet's surface in 2012. Several hundred people make the Mars Exploration Rover Mission possible, though three women working at NASA's Jet Propulsion Lab in California – Courtney O'Connor, Stephanie L. Smith, and Veronica McGregor – are behind Curiosity's popular social media presence.

Next Nature Network

Founded in 2011 in Amsterdam, the Netherlands, Next Nature Network is a 21st century nature organisation that focuses on the connections between biology and technology in its events, exhibitions, publications, and products, in which the two fields converge. Next Nature aims to spread awareness about the changing relationship between people, nature, and technology and to strengthen the connections between the biosphere and the technosphere. The Next Nature philosophy inverts the idea that nature and technology are opposed; rather, humans are so surrounded by technology that they should aim to live in harmony with it as a "next nature."

Niantic, Inc.

The American software developer was founded as an internal start-up at Google in 2010 under the name Niantic Labs. The company, independent since 2015, is led by John Hanke, who has come under scrutiny by data privacy advocates for his work on Google Earth and Street View. The company is best known for its development of Pokémon Go for Nintendo, a Japanese manufacturer of videogames and consoles. Nintendo, founded in 1889 as a manufacturer of playing cards, has also produced the Game Boy and Wii gaming consoles as well as numerous best-selling games, such as Super Mario and Pokémon.

Nintendo

Founded in 1889 as a playing card and toy manufacturer, the Japanese company Nintendo began specialising in video games and consoles in the early 1970s. The 1985 title R.O.B., a fusion of toy robot and video game, thus represents a combination of these two aspects of the company's history. After a drop in sales in the United States, R.O.B. represented a push by the company to introduce a novelty in the video game market that went well beyond the scope of conventional virtual video games.

Johanna Pichlbauer and Mia Meusburger

Johanna Pichlbauer (b. 1989 in Graz, Austria) and Mia Meusburger (b. 1995 in Bregenz, Austria) are design students at the Department of Industrial Design at the University of Applied Arts in Vienna. In 2014, they collaborated in creating the speculative design project Vienna Summer Scouts where they explored ways of measuring the emotional potential of a city by means of small digital sensors placed across the city. This work received international design media attention.

Eric Pickersgill

(b. 1986 in Homestead, Florida, USA) is an American photographer and artist. He received a Master of Fine Arts degree at The University of North Carolina at Chapel Hill in 2015, and a Bachelor of Fine Arts degree with a concentration in Fine Art Photography from Columbia College Chicago in 2011. His works are often about photography, exploring the psychological and social effects that cameras and their artifacts have on individuals and societies as a whole.

Joseph Popper

(b. 1986, London, Great Britain) studied Design Interactions at the Royal College of Art in London. His work deals with space travel and other human technological endeavours by imagining future narratives and simulating fictional experiences through film, design, and architecture. Popper is interested in science fiction as a means of approaching unknown frontiers.

Alex Proyas

Director Alexander Proyas (b. 1963 in Alexandria, Egypt) was raised in Australia and began his career in Los Angeles in the science fiction and fantasy genre. His 1994 fantasy film *The Crow* was a critical success, but he did not return to the silver screen until 2004, when the blockbuster *I, Robot* starring Will Smith was released. Known for highly stylised production design, Proyas released the big-budget film *Gods of Egypt* in 2016.

Gerard Ralló

(b. 1984 in Barcelona, Spain) is a Spanish designer and technologist based in Tokyo, Japan. He is a graduate of the Design Interactions programme at the Royal College of Art in London. Ralló's conceptual and speculative design projects question the use and impact of technology in society. His works appeared in books and exhibitions around the world, including at the MoMA in New York.

Alexander Reben

(b. 1985 in New York, USA) is an artist, lecturer, entrepreneur, and roboticist. Despite coming from a background in engineering, mathematics, and robotics, he also embraces psychology, philosophy, and design in his work. This fusion is a must, because since completing his MA at the MIT Media Lab in 2010, Reben has created numerous projects and installations that raise provocative questions about what it is to be human in our technologically-driven world – a dilemma that should not be underestimated.

Robo Technologies

Founded in 2013 by Rustem Akishbekov, Anna Iarotska, and Yuri Levin, the Vienna-based start-up has developed a robotic toy. Using electronic bricks, children can build their own robots and control them using an iOS or Android app. In 2014, the team was part of the hardware accelerator HAX and earned the "Robot of the Year" award from Festo and the "Austrian Start-up of the Year" award.

Robotlab

The collective was founded in 2002 by Matthias Gommel, Martina Haitz, and Jan Zappe at the ZKM Centre for Art and Media in Karlsruhe, Germany. They are an independent group of artists who create installations and performances with industrial robots that question the relationship between human and machine and propose different modes of interaction with robots in a cultural context.

Gene Roddenberry

(b. 1921, El Paso, Texas, USA, d. 1991, Santa Monica, California, USA) was an American television screenwriter and producer. He is best known for creating the original *Star Trek* television series, which generated six feature-length films with the original cast. His influence on popular culture was enormous; *Star Trek* became the first TV series to have an episode preserved in the Smithsonian Institution and NASA named one of its space shuttles Enterprise after the show's famous starship.

Rafaël Rozendaal

The Dutch-Brazilian artist Rafaël Rozendaal (b. 1980 in Amsterdam, Netherlands) graduated from the Art Academy in Maastricht in 2002. The artist's focus is certainly online – with work like the 2007 interactive website *jellowtime.com,* which has already collected millions of viewers – yet his pieces often transcend the screen. His recent work includes wall paintings developed from social media posts and geometrical tapestries created from the pixilated wireframes of websites. Rozendaal lives and works in New York.

Philipp Schmitt, Stephan Bogner, and Jonas Voigt

Philipp Schmitt (b. 1993 in Würzburg, Germany), Stephan Bogner (b. 1993 in Freyung, Germany), and Jonas Voigt (b. 1992 in Hof, Germany) are multi-disciplinary designers. They graduated from the Interaction Design programme at the University of Design in Schwäbisch Gmünd, Germany, in 2016. The team worked together on the speculative design project Raising Robotic Natives, for which they imagined new objects to help the next generation to interact with robots in a domestic setting. The project was widely reported on by design media across the world.

Jake Schreier

(b. 1980 in Berkeley, California, USA) is a film director and producer. A graduate of New York University's Tisch School of the Arts and a founding member of the Brooklyn-based filmmaking collective Waverly Films, Schreier began his career making short films and music videos. He joined Park Pictures in 2006, releasing his first feature film, *Robot & Frank,* in 2012. The film won the Sundance Film Festival's award for best feature film on science or technology.

Senseable City Lab

The Sense*able* City Lab at the Massachusetts Institute of Technology (MIT) in Boston, USA, is a research laboratory, which investigates and critically analyses the current transformation of the city through digital networks and digital information. In doing so, new approaches for exploration and interpretation of our constructed surroundings are generated. Managed by the Italian architect Carlo Ratti, the lab is omni-disciplinary: Designers, planners, engineers, physicists, biologists and social scientists are integrated in and cooperate with industry, local government, individuals, and communities.

Masamune Shirow

(b. 1961 in Kobe, Japan) is the pen name of the best selling manga creator Masanori Ota. His most famous work is *The Ghost in a Shell* manga sequence, which was adapted for anime movies and TV series as well as video games and toys. He studied at the Osaka University of Arts, where he currently teaches at the Oil Painting Department.

Takanori Shibata

(b. 1967 in Nanto, Japan) is an engineer at the National Institute of Advanced Industrial Science and Technology in Japan. He created Paro, a therapeutic robot developed to interact with humans and to be used in the treatment of mental health patients and the elderly. Among his many awards, Dr. Shibata received the Good Design Award, Japan's most prestigious design award, and the Guinness World Record for "The most therapeutic robot" in 2002.

Robert R. Snody

(b. 1898, New York, USA, d. 1982, Los Angeles, California, USA) was a film director, writer and producer. Films include *Rigoletto Blues* (1941), *Di que me quieres* (1939), and *The Middleton Family at the New York World's Fair* (1939), the latter being Snody's most notable work as a director and screenwriter. In 2012, the National Film Registry acquired the film for its cultural, historical, and aesthetical significance in American filmmaking.

Hajime Sorayama

(b. 1947 in Ehime, Japan) is a Tokyo-based illustrator known for his hyper-realistic depictions of sensual female cyborgs, which may be regarded as the pin-ups of the future. To achieve this realism he works by hand with acrylic paint and airbrush. Sorayama, whose art is famous all over the world, has worked for major companies and franchises including Nike, Lucasfilm, Playboy, Marvel Comics, Star Trek, Disney Media, and the Sony Corporation. He designed the original concept of AIBO, Sony's canine robot, in 1999, winning both the Media Arts Festival Grand Prize and the Good Design Grand Prize, one of the most prestigious design awards in Japan, for this work.

Ismael Soto

(b. 1988 in San Diego, California, USA) is a San Francisco-based architectural designer. His work explores the spatial implications of responsive robotic environments and emerging digital technologies in architecture. He studied under digital design pioneer Greg Lynn at the University of California, Los Angeles. He has worked for Zaha Hadid Architects in London and Coop Himmelb(l)au in Vienna. Currently he works at Skidmore, Owings & Merrill LLP (SOM).

Andrew Stanton

(b. 1965 in Rockport, Massachusetts, USA) is a filmmaker based at Pixar Animation Studios. He studied character animation at The California Institute of the Arts (CalArts) in Los Angeles. Stanton joined Pixar in 1990, becoming the company's second animator. He wrote and directed *Finding Nemo* (2003) and *WALL-E* (2008), and *Finding Dory* (2016) for the studio. He also co-wrote all three *Toy Story* films and *Monsters, Inc.* (2001).

Starship Technologies

(founded in 2014) is a delivery robots start-up based in London and Estonia. Launched by two Skype co-founders, Ahti Heinla and Janus Friis, the company seeks to make local delivery systems smarter by building a fleet of low-cost, time-efficient, and environmentally safe self-driving robots.

Kim Swift and Erik Wolpaw

Kim Swift (b. 1983, USA) is one of the best-known figures among the young generation of game designers. After graduating from the DigiPen Institute of Technology in Washington, she worked together with some fellow students to develop a game that operated along similar lines to *Portal*. This brought her to the attention of the gaming company Valve, who hired her to work on the concept for *Portal* and numerous other games. In 2007, she received the Game Developers Choice Award together with fellow *Portal* author Erik Wolpaw (b. 1967, USA).

Jacques Tati

(b. 1908 in Le Pecq, France, d. 1982 in Paris, France) was a French filmmaker and actor who gained renown for comedy films in which he portrayed people in conflict with the mechanised modern world. In most of his films he played his signature character Monsieur Hulot, a cranky, pipe-smoking fellow with a curious, innocent nature. He wrote and starred in all six of the feature films that he directed. Wider acclaim came in 1959 with an Oscar for *Mon Oncle*.

Superflux

The Anglo-Indian research design and foresight company was founded in 2009 by Anab Jain and Jon Ardern. Both founders hold an MA in Design Interactions from the Royal College of Art. The studio has expertise in strategic and design-led futures, speculative design, and technology innovation. They have worked with major clients such as Microsoft and Samsung and have exhibited at the MoMA in New York, the National Museum of China, and the V&A in London.

Takara and Hasbro

The Japanese toy manufacturer Takara Tomy was formed in 2006 as the result of the fusion of the plastics producer Takara (founded in 1955) and the toy manufacturer Tomy (founded in 1924). Together with the American toy and writing materials manufacturer Hasbro (founded in 1923), Takara Tomy produces the Transformer series of action figures. According to the company's own statistics, its best-known products also include the Beyblade line of spinning tops. Hasbro, too, is a global company whose products include bestselling toys and games such as My Little Pony and Monopoly.

Kibwe Tavares

(b. 1983 in London, Great Britain) is a British architect and filmmaker. A co-founder of the Factory Fifteen studio, Tavares combines his architect's vision with storytelling and animation to create futuristic 3D live action/animated films. He studied at the Bartlett School of Architecture. His graduation film, *Robots of Brixton,* won the Special Jury Prize at the Sundance Film Festival. In 2012, he was on Fast Company's "100 Most Creative People in Business" list, and received a prestigious TED fellowship.

Osamu Tezuka

(b. 1928 in Osaka, Japan, d. 1989 in Tokyo, Japan) was born into a liberal home and already as a schoolboy put his imaginative talent to work by drawing comics. He had an important influence on the media socialisation of the post-war generation in Japan and his new comic style paved the way for the later emergence and popularity of manga and anime. This earned him nicknames like "the father of manga" or "the Japanese Walt Disney".

Universal Everything

The digital art and design collective Universal Everything (based in Sheffield, Great Britain) was founded by the artist and graphic designer Mike Pyke in 2004. The studio incorporates cutting-edge technology into diverse commissions such as "Energy", the moving image identity of the 2012 London Olympic Games and *Presence,* a mesmerizing video artwork based on the dancing human form made in collaboration with choreographer Benjamin Millepied and the LA Dance Project. Universal Everything has exhibited worldwide, including solo shows at the V&A, the Science Museum in London and the Sydney Opera House.

Dirk Vander Kooij

(b. 1983, Purmerend, Netherlands) is a graduate of the Design Academy Eindhoven. Combining tradition and technology in the form of craftsmanship and digital robotic technology, he explores the limits and processes of rapid prototyping. In 2011, he won the Dutch Design Award for his graduation project, the *Endless* collection.

Vecna Technologies

Vecna is a U.S. technology company located in Cambridge, Massachusetts, USA with a focus on the healthcare and logistics sectors. Founded in 1998 by Debbie and Daniel Theobald, the company initially worked exclusively in IT consultancy and systems integration within the U.S. Military Health System and the Department of Veterans Affairs. Today it mainly produces high-tech medical products.

Richard Vijgen

(b. 1982 in Tilburg, Netherlands) is a designer who investigates new strategies for contemporary information culture. He creates projects that connect the digital realm with physical or social space. Using Big Data, code, pixels, and 3D printers, Vijgen produces interactive data visualisations and data installations ranging in scale from the microscopic to the monumental. His work has been exhibited at the ZKM Centre for Art and Media in Karlsruhe and the Los Angeles County Museum of Art (LACMA), amongst others.

The Wachowskis

The 1999 film *The Matrix* enjoyed immense success all over the world. It is the second film by siblings Lana (formerly Laurence, b. 1965 in Chicago, Illinois, USA) and Lilly Wachowski (formerly Andrew, b. 1967 in Chicago, Illinois, USA). Before taking up filmmaking, script writing, and producing, the pair worked as comic authors. They were awarded numerous prizes for their work on *The Matrix.*

Fred Wilcox

(b. 1907 in Tazewell, Virginia, USA, d. 1964 in Beverly Hills, California, USA) was an American film director. He began his career as a publicist for Metro-Goldwyn-Mayer (MGM), where he worked for many years and eventually rose to the position of film director in the studio. His most famous works include the classic family film *Lassie Come Home* and its two sequels, *The Courage of Lassie* and *Hills of Home,* in the 1940s as well as the seminal science fiction movie *Forbidden Planet* in 1956.

Vogt + Weizenegger

Oliver Vogt (b. 1966 in Essen, Germany) and Hermann Weizenegger (b. 1963 in Kempten, Germany) founded their design studio Vogt + Weizenegger (V+W) in 1993 in Berlin. Their method focused on systematic design which shaped not only objects, but also contexts. Alvin Toffler's concept of a productive consumer, the "prosumer", whereby the consumer becomes an active producer, greatly influenced their work. They were exhibited internationally and won several prizes, such as the Red Dot, the iF Design Award, and Design Plus. V+W was dissolved in 2008.

Anouk Wipprecht

(b. 1985 in Purmerend, Netherlands) describes herself on her website as a designer, an engineer, and a curator. She studied interaction design alongside fashion design and has already won many prizes for her "Fashion-Tech" creations – fashion that uses integrated technology. She often collaborates with major companies. The technology for her *Spider Dress,* for instance, was made by Intel-Edison. Other designs have resulted from collaborations with Google, Microsoft, and Audi. She divides her time between Amsterdam, San Francisco, and Los Angeles.

Women's Tech (WoTech)

Led by Therese Izay Kirongozi, WoTech is an association of female engineers in the Democratic Republic of the Congo. Beginning in 2013, the group began producing large, aluminum-plated traffic robots to be sold to the Congolese government. The campaign received much media attention on account of the simplicity – but general effectiveness – of both the robot's design and its adherence to pop culture stereotypes of how a robot should look.

Master Xianfan / Master Xuecheng

The Buddhist monk character Xian'er originally appeared in a book written by a Buddhist monk named Master Xuecheng (b. 1966 in Xianyou, China). The character was later brought to life – first in illustrations for the book, and later as a robot by the same name – by Xuecheng's student, Master Xianfan (b. 1987, China). After the latter completed his art studies, he became a monk at the Longquan Monastery near Beijing in 2011, where he now serves as abbot.

Steve Worswick

(b. 1970 in Wakefield, Yorkshire, Great Britain) is a British IT consultant. He started programming chatbots as a way of attracting more visitors to his dance music website. In 2004, he was commissioned by an American games company to write Mitsuku, a new chatbot. Mitsuku won the top prize in the 2013 and 2916 Loebner contest – where chatbots, artificially intelligent programs, are tested to see how convincingly they can simulate a human conversation. Worswick continues to refine Mitsuku's code.

Oren Zuckerman

(b. 1970 in Jerusalem, Israel) is the founder and co-director of the Media Innovation Lab (miLAB) of the Interdisciplinary Center (IDC) Herzliya, Israel, where he also holds a chair for interactive communication. Zuckerman, who studied at the renowned MIT (Massachusetts Institute of Technology, Boston, USA), focuses his research on the interface of interactive technologies and human behaviour. In particular, he investigates physical and digital experiences of interaction.

ACKNOWLEDGEMENTS

Robotics, informatics and information technology, artificial intelligence research, media theory and media art are complex subjects. The curatorial team for *Hello, Robot. Design between Human and Machine* is extremely grateful to a large number of experts without whose support neither the exhibition nor this publication would have come about.

First and foremost, therefore, we would like to thank the consultants to *Hello, Robot.* who patiently answered our numerous questions:

Bruce Sterling, science fiction author, network activist, design thinker, and cyber-space theorist
Carlo Ratti, designer, architect, urban planner, founder of Carlo Ratti Associati, Turin, Italy, and Director of Sense*able* City Lab at the Massachusetts Institute of Technology (MIT) in Cambridge, Massachusetts, USA
Gesche Joost, Professor for Interaction Design and Media, head of the Design Research Lab, Universität der Künste, Berlin, Germany
Sabine Himmelsbach, art historian and curator, Director Haus der elektronischen Künste Basel, Switzerland
Paul Feigelfeld, scientific coordinator of the Digital Cultures Research Lab at the Centre for Digital Cultures at Leuphana Universität Lüneburg, Germany

In the name of the three participating museums, the curatorial team would also like to thank the following experts for their help and support:

Pieter Ballon, Professor of Communications Science at Vrije Universiteit, Brussels, Director of imec SMIT-VUB, Brussels, Belgium
Giulia Bini, co-curator of the exhibition *GLOBALE: Exo-Evolution,* ZKM | Zentrum für Kunst und Medien Karlsruhe, Germany
Sigrid Brell-Cokcan and Johannes Braumann, architects and founders of the Association for Robots in Architecture, Vienna, Austria
Mark Coeckelbergh, Professor of Philosophy of Media and Technology at the University of Vienna, Austria; Vice-President of the Society for Philosophy and Technology; Professor of Technology and Social Responsibility at De Montfort University, UK; Member of the Technical Expertise Committee at the Foundation for Responsible Robotics
Hendrik Dacquin, CEO and digital interaction design expert, Small Town Heroes, Gent, Belgium
Walter De Brouwere, CEO, doc.ai, San Francisco, California, USA
Bart De Waele, CEO and coach, Wijs, Gent, Belgium
Wim De Waele, CEO and mob leader, Eggsplore, Brussels, Belgium
Wim Forceville, VR and human interaction design expert, Gent, Belgium
Karin Gimmi, curator of the exhibition *Please Touch!*, Museum für Gestaltung Zürich – Schaudepot, Switzerland
Fabian Hemmert, design researcher, Professor of Interface and User Experience Design, Bergische Universität Wuppertal, formerly a researcher at the Design Research Lab, Universität der Künste Berlin, Germany
Margarete Jahrmann, media artist and Professor of Game Design, Zürcher Hochschule der Künste, Zurich, Switzerland, and lecturer at the Universität für angewandte Kunst, Vienna, Austria
Kasper Jordaens, Tru Lefevre, Tim Rootsaert, Bas Baccarne, Sarrah Logge, Change Agents at imec, Belgium
Kevin Kelly, author and publisher, San Francisco, California, USA
Sabiha Keyif, co-curator of the exhibition *GLOBALE: Exo-Evolution,* ZKM | Zentrum für Kunst und Medien Karlsruhe, Germany
Prem Krishnamurthy, designer, curator, and author, founder of the design firm Project Projects and the exhibition space P! in New York, USA, and son of Bala Krishnamurthy, the robotics pioneer who collaborated with the industrial robotics legend Joseph Engelberger in the 1970s

Steven Latré, Assistant Professor, Universiteit Antwerpen, imec, Antwerp, Belgium
Jeroen Lemaire, CEO, In The Pocket, Gent, Belgium
Greg Lynn, architect and designer, Professor of Architecture at the Universität für angewandte Kunst, Vienna, Austria, and guest professor at the UCLA School of the Arts and Architecture, Los Angeles, California, and Yale School of Architecture, New Haven, Connecticut, USA
Erik Mannens, Professor at IDLab, CTO DataScience at imec and Research Manager at Universiteit Gent, Gent, Belgium
Erich Prem, computer scientist and former researcher at the Österreichisches Forschungsinstitut für Artificial Intelligence (ÖFAI) and at the AI Lab of the Massachusetts Institute of Technology (MIT), Cambridge, Massachusetts, USA, CEO, eutema GmbH, Vienna, Austria
Maria Pruckner, cybernetics specialist and organisation consultant, CEO, InForMent – Strategie, Struktur & Kultur für das Meistern hochdynamischer Komplexität, Vienna, Austria
Christian Rohner, exhibition curator, Museum für Kommunikation, Bern, Switzerland
Matthias Scheutz, Professor of Cognitive Science and Computer Science, Bernard M. Gordon Senior Faculty Fellow Director, Human-Robot Interaction Laboratory, Department of Computer Science at Tufts University, Medford, Massachusetts, USA
Christian Stadelmann, co-curator of the exhibition *Robots. Men and Machine?* and head of the department Alltag & Umwelt, Technisches Museum, Austria
Gerfried Stocker, media artist, Artistic Director of Ars Electronica Center, Linz, Austria
Robert Trappl, cyberneticist, Professor and Director of the Österreichisches Forschungsinstitut für Artificial Intelligence (ÖFAI), Vienna, Austria, President of the International Academy for Systems and Cybernetic Sciences
Gerhard Tröster, Professor of Electronics, Director of the "Digital Systems & Wearable Computing" division at the Institut für Elektronik der ETH Zürich, Zurich, Switzerland
Steven Van Belleghem and Peter Hinssen, writers and network nodes, nexxworks, Gent, Belgium
Dirk Van Welden, game designer, artist, and CEO, I-illusions, Brussels, Belgium
Bram Vanderborght, Professor of Robotics, Vrije Universiteit Brussel, Brussels, Belgium
Peter Weibel, artist, curator, art and media theorist, Director of ZKM | Zentrum für Kunst und Medien Karlsruhe, Germany

Amelie Klein would also like to thank Erika Pinner, her indefatigable assistant curator, without whom neither the exhibition nor the catalogue would have been possible; Thomas Geisler for a fruitful exchange of ideas stretching over many months and for many years of friendship; Stefani Fricker (head of the Technical Department at Vitra Design Museum) and Nathalie Opris (Media Technology at Vitra Design Museum), Valerie Hess, Raphael Höglhammer, and Pia Hönges (Emyl) as well as Veit Grünert and Chris Rehberger (Double Standards) for once again staying the course; Viviane Stappmanns and Yvonne Radecker for comfort and support of many kinds; Tanja Cunz for her support and Jolanthe Kugler for "lending" her; and all other colleagues at the Vitra Design Museum without whom *Hello, Robot. Design between Human and Machine* would not have happened.

IMPRINT

This catalogue is published in conjunction with the exhibition *Hello, Robot. Design between Human and Machine.*

The exhibition *Hello, Robot. Design between Human and Machine* is a coproduction of the Vitra Design Museum, the MAK – Austrian Museum of Applied Arts / Contemporary Art and the Design museum Gent.

Vitra Design Museum: 11 February to 14 May 2017
MAK – Austrian Museum of Applied Arts / Contemporary Art: 21 June to 1 October 2017
Design museum Gent: 27 October 2017 to 15 April 2018
Gewerbemuseum Winterthur: 12 May to 4 November 2018
A number of other venues are planned.

Catalogue

Editors: Mateo Kries, Christoph Thun-Hohenstein, Amelie Klein
Overall editorial management: Amelie Klein, Erika Pinner, Tina Thiel
Editorial management essay Rosi Braidotti: Marlies Wirth
Authors: Daniele Belleri, Rosi Braidotti, Fredo De Smet, Christoph Engemann, Paul Feigelfeld, Thomas Geisler, Lea Hilsemer (LH), Gesche Joost, Amelie Klein, Olivia Parkes (OP), Erika Pinner (EP), Carlo Ratti, Aline Lara Rezende (AR), Bruce Sterling, Tina Thiel (TT), Marlies Wirth
Cover illustration: Christoph Niemann
Graphic design: Chris Rehberger, Veit Grünert, Double Standards, Berlin
Layout algorithm programming: Timo Rychert
Project coordination for translation and copyediting: Tradukas GbR
Translations: English–German, German–English: Tradukas GbR; Wortreich/Jana Güttler, Maria Wokurka; Andrea Schellner
Essay Marlies Wirth: Wortreich/Jana Güttler, Maria Wokurka
Essay Bruce Sterling: Tradukas GbR, Jörg Blumtritt
Dutch–English: Lisa Holden
Image rights: Erika Pinner, Tanja Cunz, Eboa Itondo
Lithography: Markus Bocher, GZD Media GmbH
Production manager: Stefanie Krippendorff
Distribution: Stefanie Krippendorff, Irma Hager
Printing: DZA Druckerei zu Altenburg GmbH
Typefaces: Rockwell Std and Garamond Premier Pro
Paper: Arctic Volume HighWhite, 130g/qm

The German National Library lists this publication in the German National Bibliography; detailed bibliographical data can be accessed under http://dnb.dnb.de

ISBN 978-3-945852-10-1
(German edition)

ISBN 978-3-945852-11-8
(English edition)

Exhibition

Directors Vitra Design Museum: Mateo Kries, Marc Zehntner
Directors MAK – Austrian Museum of Applied Arts / Contemporary Art: Christoph Thun-Hohenstein (General Director and Artistic Director), Teresa Mitterlehner-Marchesani (CFO)
Curators: Amelie Klein, Thomas Geisler, Marlies Wirth
Advising curator: Fredo De Smet
Assistant curator: Erika Pinner
Project management: Amelie Klein, Erika Pinner, Sabrina Handler
Exhibition design: emyl, Basel
Exhibition graphics: Hug & Eberlein, Leipzig, Basel
Multimedia agency: nous Wissensmanagement, Vienna, Denver, Dubai
Technical direction: Stefani Fricker
Installations: Michael Simolka, Patrick Maier-Blanc, Manfred Utz
Media technology: Nathalie Opris, Adam Bagnowski
Conservation: Susanne Graner, Lisa Burkart, Grazyna Ubik
Registrar: Isabel Serbeto, Bettina Besler-Slawik, Boguslaw Ubik-Perski
Press and public relations: Iris Mickein, Philipp Hindahl, Carolin Schweizer
Partnerships: Yvonne Radecker
Exhibition tour: Reiner Packeiser, Isabel Serbeto
Accompanying programme: Sarah Kingston, Lisa Nüsseler
Education & visitor experience: Noëmi Preisig, Johanna Horde
Visitor services: Annika Schlozer, Julia Wichmann
Products: Johannes Heinzmann

Funded by

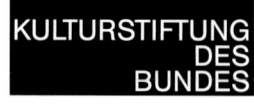

Global sponsor   Sponsor   Supported by